THE FARM
IN JONES VALLEY

BY: RAYMOND B. JONES

First Edition

Dedication

To my mother, Sarah Elizabeth Jones ("Miss Betty"), whose pioneer spirit and perseverance allowed our family to retain the farm in Jones Valley during the WWII years.

To my wife, Elizabeth Mercer Jones ("Miss Libby"), who has been an encourager and the inspiration to write this book. She is the one who deserves as much credit as anyone for the farm's survival and is a godly influence on all those that tread its fields.

Contents

List of Illustrations

Acknowledgements

Carl T. Jones – My father, who along with his brother Edwin, envisioned the potential for the farm and the one who instilled in me the love of the land.

Sarah Elizabeth Jones – My mother, who influenced me to love raising animals and provided a wonderful example of good management.

Elizabeth M. Jones – My partner in life for almost 50 years. She was the main encourager for writing this book and the one who has given my life its real meaning.

Larkin and Amanda Battle – My sometimes "foster" parents during the war.

John K. Everson – My friend who guided me in managing the farm in a masterful way even though he only had a second grade education.

Lisa Yokley – My daughter, who deserves credit for editing this book, correcting my sometimes bad English and for encouraging me to finish.

Raymond B. Jones, Jr. – My son, for correcting and re-shaping some of the later year chapters.

Carolyn Moses – My wonderful stenographer and decipherer of my poor handwriting. Without her this book would have been most difficult.

Bernice Limbaugh Huie – Former HHS classmate and friend who was most kind to assist in the editing process.

Mark Yokley – My son-in-law who provided wonderful guidance and technical expertise for the final printing of this book.

Introduction

This book is a history of G. W. Jones & Sons Farm in Huntsville, Alabama. This farm, at the writing of this book (2009) is thought to be the largest operating farm in the United States completely bounded by a major city. The farm had its beginning with the Jones family in 1939 and has continued as an agricultural enterprise to this time under the family ownership.

Someone has said that "farmers are born and are not manufactured". The most successful farmers are usually the ones that were raised on a farm. Farming is more of an art than a science. There are so many unknown variables in farming that even the best laid plans are usually interrupted by the weather, markets, labor, equipment breakdowns, etc. The farmer must "juggle" a lot of these variables to get things done in a timely manner. If all of my business endeavors, i.e., engineering, banking, real estate, insurance, abstracting, land surveying and coal mining were rolled into one, they would not approach the difficulty associated with farming. Those who till the soil or raise livestock are a special breed. They are usually optimistic as they work hand in hand with the Creator and see His miracles performed each day. Successful farmers are good stewards of the land as well as their crops and livestock. Only 1.8% of our nation's population is still farming today compared to 38% in the mid 1950's. Our nation owes a great debt of gratitude to its farmers who produce food and fiber for the other 98% as well as millions around the world.

Probably the most remarkable fact about the farm is its survival over difficult obstacles these last 70+ years. Wars, money problems, changes of leadership, deaths, weather and the encroachment of a growing city are just some of the obstacles that have been overcome during these last seven decades. The farm has survived the test of time, events, and difficulties and yet remains, operationally at least, similar to any large farm in a rural setting.

Thus the history of "The Farm" and some of its people and their stories are written in this book. Space does not allow me to record many of the stories, people and events that should be recorded. I sincerely wish this were possible and that I possessed the penmanship to properly relay the unique story of this wonderful tract of land located in God's country of North Alabama that our family simply calls, "The Home Place."

Raymond B. Jones

2009

"There is no substitute for leadership."

- Carl T. Jones

THE FARM IN JONES VALLEY

PART I

Chapter 1 - Knoxville and Chattanooga

The face of the doctor seemed to bear bittersweet news for the young father in the waiting room at St. Mary's hospital in Knoxville, Tennessee. The baby, a 7 lb 1 oz boy, was fine but the mother was not doing well. She was hemorrhaging and efforts to stop the bleeding had been mostly unsuccessful.

The date was March 23rd, 1935 and even though medical procedures were somewhat rudimentary, the doctors and staff worked feverishly to save this young mother's life. Additional doctors and staff were called in but it was at most a waiting game when nothing seemed to work.

The young parents were Carl and Betty Jones who had recently moved to Knoxville. Carl was a civil engineer, then working for the U.S. Forest service. He had left his home in Huntsville, Alabama because on the heels of the depression, there was no work even though his family owned and operated a consulting engineering business. Their engineering business in 1935 consisted primarily of land surveying services, abstract title work and an occasional engineering design project.

Work and other matters seemed unimportant to Carl at this moment in time. His thoughts were of his wife and her condition. Hours passed slowly and each report seemed to be the same – no change. Day two came and went with no sleep for Carl as he kept vigil over the situation. He had refused to even look at his newborn son without Betty to share in that experience and she was too weak from the loss of blood. In desperation, the hospital staff lowered the head of her bed until she was practically "standing" on her head in order to keep blood and oxygen flowing to her brain. Blood transfusions seemed to have little effect and Betty's condition was very serious.

Carl's sister Pauline drove up from Huntsville to try to comfort Carl and provide some relief so he could sleep. Carl would have none of it and stayed by Betty's bedside for five days at which time she began to recover. The bleeding slowly stopped and she asked to see her newborn son. Both mother and father shared the sight of the baby boy together for the first time on day five of his life. They named him Raymond Bryant Jones. Raymond was the name of Carl's brother who had died almost two years earlier and Bryant was Betty's maiden name. For the first time in six days, Carl slept, finally knowing that both his wife and son were improving. Ten days later Betty and Ray, as they would call their new son, were discharged and returned to their apartment. Betty continued to recover from this difficult childbirth event and even though she would later bear two additional children, the birth of her first child was the most traumatic.

Carl, Betty and new son remained in Knoxville only a short time with the U.S. Forest Service. An opportunity presented itself for Carl to work as an engineer in Chattanooga, Tennessee for a private consulting engineering firm. Chattanooga was 100 miles closer to Huntsville and his family. Carl was an officer in the Alabama National Guard and had been commuting to Monday night drills each week from Knoxville. Traveling 400 miles round trip each week to guard drills in the 1930's was in itself a feat, not to mention job and family strains that come to all young families. The Chattanooga move cut the travel time to National Guard drills in half which was a great relief.

In 1936 the Carl Jones family happily moved into a home on Woodvale Avenue in the Brainerd area of Chattanooga, Tennessee. Carl was working in his chosen profession as an engineer and Betty was busy as a homemaker and mother. Life was good for their family and they quickly made lifelong friends and established themselves in the community. In 1939 a second child, a girl, was born and they named her Francis Elizabeth. Francis, for a Huntsville friend of long standing, Francis Davis, and Elizabeth after Betty whose name was Sarah Elizabeth. I was four when my new sister came home from the hospital. Mother told me that my only comment about her was "Aw, she can't play." I eventually named the bright-eyed, beautiful little sister of mine "Betsy" and it stuck for all of her life.

The Carl Jones family of four seemed to be settling into the fabric of the community. They made a lifelong friendship with a family across the street by the name of Fundinger. They were a fun loving family of six, three girls and a boy. One of the girls, Lois, and my mother were close and remained lifelong friends. Carl was active in engineering circles in Chattanooga and the family attended the Brainerd Church of Christ. Also at this time, Carl and his brother Ed were raising bird dogs and I had a pet rabbit named "Foozy".

Little did we know that in 1938 we would receive a call from Carl's brother Ed outlining a business venture that would change the course of our lives and many other lives for the future.

Timing is Everything

There is a tide in the affairs of men,

Which, taken at the flood, leads on to fortune;

Omitted, all the voyage of their life

Is bound in shallows and in miseries.

On such a full sea are we now afloat;

And we must take the current when it serves,

Or lose our ventures.

Shakespeare

My father, Carl Jones, was a man who was always at least 10 years ahead of his time. I'm sure the decision didn't come easy but maybe he felt that, even though early in his career, this was the "tide in the affairs of men" for this venture. The venture was a real risk taking one for anyone but especially for a young family. The plan involved moving back to Huntsville, trying to revive the family engineering business and going heavily in debt to purchase a worn out farm south of the town of Huntsville which then had a population of 12,000. The thinking was that if the engineering business couldn't support the families, then they would farm for a living.

"The Farm" as it would become known over the years had been on the market for four years. It was not well served by roads and had no utilities, fencing or infrastructure. The only buildings were a mule barn built in 1915 and an abandoned house built by slave labor in 1823 with bricks made in the yard. The proposition was not very inviting but nevertheless the brothers took the risk and the Carl Jones family moved to the farm in 1939. The 2,500 acre farm is located in an enclosed valley southeast of Huntsville that extends from the southern slope of Monte Sano Mountain south to the Tennessee River. The valley is drained by Aldridge Creek and is scalloped on each side by several deep hollows and tree lined slopes. The earliest inhabitants were the Cherokee Indians and evidence of their occupation in the form of artifacts surface each time the land is tilled. Arrowheads, spear points, knives, bone needles and pottery have all been found over the years on the farm. Later, white inhabitants came in the early 1800's. The very first Caucasian settler in Madison County was my great, great grandfather, Isaac Criner who settled here in 1805. As far as we know he never visited the Aldridge Creek Valley. Isaac Criner made his home near New Market, Alabama on Mountain Fork Creek and lived there until he died in 1876 after the Civil War.

The first family to actually own part of the farm was the Drake family. John Drake was a member of this family and is buried in a small cemetery on Garth Road across from where the Mayfair Church of Christ is now located. John Drake was a Revolutionary War Soldier and died in the main dwelling on the farm in 1839.

Later owners and the family we purchased the farm from was the Garth family. Garth Road and Drake Avenue are still named for these families today. The valley was often referred to as "Garth's Hollow" when we came in 1939. One of the Garth family owners was Winston Garth and he and my father remained friends for many years after the purchase of the farm. The purchase launched more than just a business venture, it launched a way of life for our family and hundreds of faithful employees for many years to come. I'm sure other owners have cared for and loved the land in this valley, but none more than our family. The ownership remains in the Jones family today almost 70 years beyond its purchase. It is the largest active urban farm in America and a real legacy to Carl and Ed Jones who started it all.

This is a view of the farm in the 1940's looking northeast towards
Huntsville and below in 2009.

Chapter Two – Huntsville

My first trip to the farm in 1939 was in a mule drawn wagon by way of Four Mile Post Road. Whitesburg Drive was the main road leading south from Huntsville at that time and it intersected a gravel surfaced road named Four Mile Post Road which made its way east into what was known then as "Garth's Hollow." The name "Whitesburg" was derived from Whitesburg Landing on the Tennessee River which was a cotton loading station for boats transporting cotton in the previous century to New Orleans and other Mississippi River ports. Four Mile Post Road derived its name because it was four miles from the Madison County Courthouse and once had an established post marking the southern terminus for buggy races held in the 1920's. Neither road was very good but we traveled in a car until we met a wagon somewhere on Four Mile Post Road. It had been raining and the mules pulled the wagon along a route close to what is now the location of Garth Road to the farm, barn and house area.

The main farmhouse was in terrible disrepair. It was constructed over 120 years earlier and was of a "raised cottage" design. The two story structure was rectangular in shape, had a tin roof and the bottom story was submerged 1/3 of its height in the ground. I suppose this was an effort for heating and cooling the house. The exterior walls were 27 inches thick and the interior walls 18 inches thick. No one lived in the house at the time. One end of the lower floor was a farm implement storage area and the other end was a chicken coop. The house had four fire places and a small kitchen with a wood burning stove. Of course there was no water or indoor facilities. There was a pretty good well across the road from the house and a large garden out back.

The main farm house in the 1940's which was built in 1823.

Even though I was very young at the time, I remember considerable discussion evolved around our family living in this house. The 2,500 acre farm purchase had been made with a loan from Metropolitan Life Insurance Co. and the plan was for Carl's family to live and work the farm. Edwin and his family would continue to live in town and concentrate on re-building the struggling engineering business. Carl and Ed were both civil engineers with degrees from the University of Alabama. Ed was really more interested in the agricultural side of the business than Carl but had already established a

30

home on Clinton Street. The thinking was that since Carl had to move from Chattanooga anyway it would be more efficient to have him move directly to the farm.

The most daunting task for my father was to convince my mother to live in this run down home almost an hour's travel time from town. Carl's mother didn't help any because she told my mother to just say "no" to the move. Eventually Carl prevailed and assembled carpenters and other workers to renovate the house on a fast track basis while we lived in town with my grandparents, G.W. and Elvalena Jones.

Work also proceeded on improving what is now Garth Road which provided a shorter way to town. The farm came with 26 tenant families and 26 teams of mules. These sharecropper families worked the land in mostly cotton and corn. My father worked these mules and the farm hands on Garth Road with "flip scoops", grading it to a passable state. Mr. Roy Stone was the Madison County Commissioner at the time and my father got him to agree to machine grade the "flipped up" road one time when the rough grading was complete. Mr. Stone graded the road and we used this "new" road in its graded state until it was paved some 10 years later in 1949. Garth Road intersected with what is now Drake Avenue and cut travel time to town almost in half.

In addition to the road and house work, several wells were dug in order to provide water for the farmstead. After many failures, my mother got a man to "witch" a well location with a forked peach tree branch and we found a good well which served as our main water supply until a public water line was installed in the 1950's.

Still another effort that paralleled those improvements was my father's effort to convince Huntsville's Electric Utility Authority to serve the valley via the newly graded Garth Road. When we first moved to the farm we used "Aladdin" coal oil lamps. Electricity not only added light to things but power to pump water for indoor plumbing which was a great blessing for us all. With much effort and 18 hour days, my father and mother made the old house livable in a little less than six months. Little did they realize what was about to happen, because had they known that WWII was coming, I doubt they would have taken on such a risky venture and family move.

My father was the Battalion Commander of the 151st Combat Engineer National Guard Unit at the time of his move to the farm. Probably because Alabama begins with an "A", this National Guard unit was mobilized a year before war was declared. Less than six months after the farm was purchased, both Ed and Carl left for service in Alaska. Carl served in the Aleutian Islands and Ed at Nome, Alaska. My father fell in love with Alaska and its wildlife even though he was busy with his Engineer Unit building airfields and roads for which he was well qualified. My mother, a city girl raised in Memphis, Tennessee, was left to run the farm. She was the only white woman in the valley at the time but she was a good manager. She held things together during the absence of our family leaders for the duration of the war, a period of almost five years. After the war, both Ed and Carl gave her the credit for keeping the farm intact. Their military pay was used to make mortgage payments. Money from the crops, timber, livestock, molasses and other income was used for the farm operation. The instruction they gave Betty was that if they were killed in the war, she was to sell the place at the earliest opportunity.

Chapter Three – The War

The day my father left to go to war he called me, along with a black man named Larkin, to the front gate. Larkin and his family, wife Mandy and daughter Necy, came to us about the time we were moving to the farm. Larkin was a very physically strong man and proved to be an excellent manager and leader of men on the farm. Mandy was an excellent cook and was our family cook until she died after almost 30 years of being a part of our family. Necy lived in our house during the war as "protection", I suppose, even though she was still in grade school.

The dialog of this meeting between my father, Larkin and myself went something like this:

"Larkin, you know I've designated you as my representative to farm in my place which exempts you from the draft." (During WWII the government would allow a farmer to appoint a designee to farm in his stead should he be called into service.) "Yes sir" he replied. "I want you to help Miss Betty keep the farm going and be here when I come back." "Yes sir, I'm gonna do just that, Mr. Carl", Larkin said.

"Ray, I don't know how long I'll be gone and you may be a big boy before I get home, in fact you may be too big for your mother to spank before I return." "Oh, Daddy I'm sure I won't need a spanking" I said. "But if you do, Larkin is to do the spanking. Your mother will tell Larkin when and how severe. Remember now it's not Larkin that will do

the real spanking, it will be me and your mother, understand?" "Yes, sir", I agreed.

Larkin and I looked at each other kinda funny. Over the next five years he only had to spank me once or twice at mother's direction but each time he did he would say, "Now Bubba, remember this ain't me, it's your momma and daddy." I couldn't have had a better one for this task than Larkin because he loved me like a son.

When the war was over in 1945, Larkin met my father at that same front gate with the words "Well Mr. Carl, I's still here". My dad never forgot this and Larkin worked for the farm the rest of his working life and was like a member of the family.

As a young man I was having the time of my life. I was outside every minute I could manage. I had a horse, two dogs and I raised bantams, turkeys, and chickens. I hunted in the winter and fished in the summer. Work was fun. I had numerous friends, all black boys living on the farm. We swam, fished, worked and hunted together and I enjoyed it all. Work consisted of plowing cotton with mules, combining clover and oats in the spring and raising livestock. We had several livestock ponds on the farm that provided swimming and good fishing. I remember swimming with a bunch of boys one day in particular. We would paddle out to an old submerged metal water tank, stand on it a while and paddle back to the bank. On this particular occasion I remember one boy, Roosevelt Hereford, missed the submerged tank and began to drown. I swam to him but I knew better than to let him grab me as he was frantically trying to do. I dove down below him, picked him up by the legs and walked on the bottom of the pond toward the shore. I had to do this several times until we reached shallow water. Roosevelt was about gone and had turned a greenish color from lack of oxygen. Upon reaching the shore I pulled him by the feet up the bank and pounded on him until water gushed out of his mouth and nose. I didn't know how to resuscitate a drowning victim but this must have worked for in about an hour he was okay. Every time I saw him, even years later he would always say "Remember that time you saved my life?" And then we would talk about it for awhile.

Mother did the best she could to run the farm during the war years. She had the crops planted and harvested, made a garden, made molasses, and cut and marketed logs from the farm. She was an excellent manager. She tried to take advantage of the idle labor during the winter months after the crops were harvested. She would employ the men to clear land and sell the logs. During the war timber sold at excellent prices. Each family had a team of mules and they were used in the timber cutting and handling. Of course each man thought his team was the best and would bet and brag as to how much they could pull.

On one occasion, I remember we left the woods with three wagons loaded down with logs. The first wagon made it through a bad mud hole but the second one got stuck. Albert Battle was the driver of the stuck wagon and was embarrassed because he had overly bragged on his team, ole Dick and Dave. Albert proceeded to whip his team unmercifully and called them every name in the book. Ole Dick and Dave balked like some mules do and when they do balk they will not move. In desperation, Albert took some hay from the third wagon, spread it under Dick and Dave and lit it with a match.

The fire caused Dick and Dave to stomp and kick and then they pulled up just enough to catch the wagon and axle grease on fire. The wagon was destroyed and some of the logs burned and Dick and Dave had the last laugh. After that, Albert never bragged on his team again.

Mules were very important in those days before we had tractors. A fond image in my mind of that bygone era was seeing the teams coming in from the fields at quitting time. After plowing all day, each man would come to the barn riding one mule and leading the other. Many times they would be singing, glad the hot day was over. I would race down the road in order to ride the mule that was being led back to the barn and then race back to catch another. Oh how I wish I could see and hear those scenes again.

Everyone had a "mule" story that we would enjoy from time to time. One sharecropper, Will Bill Battle, said "that a mule will work for you faithfully all his life just to get to kick you once real good". Nostalgic as they were, we were all glad when John Deere and Allis Chalmers tractors came on the scene.

The 1915 vintage mule barn is still standing at the farm today and is in pretty good shape. It would house 14 teams (28 mules) in its heyday. Since it has ceased being used by mules, it has been used for fescue seed and hay storage, a pleasure and working cow horse barn, a 4-H club calf grooming facility, as well as many other uses. We don't know who built the barn but I'm sure they would be surprised that the barn has lasted almost 100 years. Whoever built it did a good job and it has been an important part of our farm operation.

Mules and the mule barn.
It was built in 1915 and pictured here in the 1940's and below in 2009.

Chapter Four – Watermelons and DDT

My father was in the Aleutian Islands in 1942 and was involved in what was later termed "The Thousand Mile War". He served with Generals Buckner, Wainwright, DeWitt, Corlett and others as our American forces plodded westward in an effort to re-take the islands of Attu and Kiska from the Japanese. Generals Wainwright and DeWitt each had Springer Spaniel dogs that they bred just prior to the Aleutian Campaign. They gave my father a female puppy from this mating and it was a grand day for me as I met the train with my mother to get this cute puppy that we named "Pooch". Little did we know that she would be the start of a bloodline of Springer Spaniel duck retrievers that would last for over 50 years.

I had a big dog named Brownie, our Springer Spaniel, Pooch and a horse named Kissie (short for the Alaskan Island of Kiska) and I was having a good time being a boy living on a farm. My sister Betsy also got a small pony about this time named Polly. Polly wasn't much to ride but she would pull a little green four wheel wagon we had around the farm. The wagon had steel wheels, no springs and a little seat up front for the driver. One day my friend Roosevelt and I hooked Polly up and took a ride up what is now Garth Road. We traveled north on the dirt and gravel road until we got near what is now Jones Valley School. Our plan was to turn east on another field road, cross Aldridge Creek and make a long loop through the farm back to the barn. When we crossed the creek, we saw a big watermelon patch on the right that belonged to Mr. Osborne. The watermelons were ripe so we stopped and ate one. Then we loaded another one or two in the wagon to take back. I don't know whether the wagon looked empty or greed set in but before many minutes passed we had loaded Polly's wagon to capacity. Now we decided to return by way of Garth Road instead of the long way around. Polly had to strain to pull two boys and the wagon full of watermelons up the hill. When we reached Garth Road, we saw Mr. Osborne coming down the road. He was still a pretty long way off so, like in the movies, we made a run for it. The scene was this; two boys, one driving and whipping Polly and one riding shotgun, making a getaway down Garth Road going back to the barn. Watermelons were bouncing out of the wagon and smashing onto the road. Polly was in a full gallop and Mr. Osborne was in distant pursuit. The getaway went pretty well until we turned into the driveway leading to the barn. The turn and the weight of our load caused the front axle to break spilling watermelons and boys all over the intersection. Polly stopped and we sheepishly stood over the wagon and watermelon carnage waiting on the certain arrival of Mr. Osborne. Momentarily he arrived and asked if we were hurt. "No sir," we replied. He helped us get the wagon out of the road and unhook Polly so we could go home. "We're sorry about your watermelons Mr. Osborne, we'll be glad to make it up to you", I finally was able to say with a trembling voice. "That's alright boys. I was a boy myself once upon a time, but if you had asked me for some watermelons I would have given you all you wanted".

I never forgot this incident which served as a good lesson in taking things that didn't belong to you. We all thought more of Mr. Osborne after this incident for the way he reacted to a couple of boys taking his watermelons.

About this same time, the insecticide DDT was introduced in the stores, particularly to those stores catering to farmers. This wonderful discovery brought relief from nagging insects, especially the common housefly. We purchased DDT dust as soon as it was available to use at the barn and around the house. We were milking three cows at the time and both the milk cows and the person doing the milking were thankful for the DDT. Mother had us also dust DDT around the back porch to help with the flies and "de-flea" the cats and dogs.

One day I was dusting DDT around the house back porch with a little hand pump duster. Betsy, who was a little over three years old, announced to me. "I want to dust the DDT". "No, Betsy, you can't dust the DDT because you're too little." "I'm not too little and I want to dust the DDT", she demanded. "Betsy, you can't dust the DDT" I replied emphatically. "WAH-WAH-WAH" she began to cry real loud. My mother heard the commotion and bounded out the door. "What's wrong darling?" she asked. Whimpering Betsy said, "Ray sprayed the DDT on me." Without a moment's hesitation, mother flew into me and there was no way to direct her from giving me a good thrashing. Betty had real small hands but she knew how to use them for disciplinary purposes. Years later after I returned home from serving in the Army, I "sassed" her one day and she let me have a sharp hand lick across my shoulder. We both laughed about it because she hurt her hand but she was right about the situation.

While she was furiously whipping me, Betsy quit crying and tugged at mother's coattail and said "Aw, mama I was just teasing." At this stage mama didn't know what to do so she laughed. Betsy didn't laugh but looked satisfied. It was too late for me since I had already received "unjust" punishment.

Mother said later that I got some of what I deserved for other occasions that went unpunished.

In any event, we all knew who was in charge and Betty carried out in an admirable way the role of the family leader, farm manager and person in responsible charge. She, along with the rest of those on the home front, prayed and longed for the day when those who served in foreign lands could return home and assume their place in society. The war had caused a sudden shift in responsibility and leadership to those left behind and they deserved as much credit for winning the war as our soldiers, sailors and airmen.

Chapter Five – Brownie

One of the most colorful characters on the farm was a sharecropper named Will Bill Battle. Will Bill and his wife Paralee were the hardest workers we had and their crop was always better than the others. In the fall, Will and Paralee could, between them, pick a bale of cotton a day. They had a big family but individually they could out work any of their children. In addition to being a good farmer, Will was always looked upon as being the philosopher of any group assembled.

I shall always remember the scene in front of our house early the next morning after my father came home from the war. Men, maybe 75 to 100, were assembled to get to see Mr. Carl. My dad had been gone almost five years and they lined up to hug and greet him on that particular morning. Some of them wept tears of joy when seeing him and all were in a festive mood. My father brought out with him an Army C-Ration for them to see. They had never seen a combat ration before and my dad opened the C-Ration carton and spread out the contents on a big rock for all of them to see. A WWII vintage C-Ration consisted of just the bare essentials to sustain a soldier in combat; a piece of dried fruit and meat, a small cake, a cigarette, toilet paper, etc. The men were very interested and questioned my dad as to the rigors of combat army life. When a lull came in the discussions, Will Bill pushed his hat back on his head and everybody knew he was going to expound on something. He said, "Now Mr. Carl, I know you are a smart man and I know the government has got some smart mens. But if all they gonna put in dar for a soldier to eat is that little piece of dried meat, fruit and cake, den they sho didn't need to put no toilet paper in there". The whole group must have laughed for five or ten minutes. Over the years, I heard my father tell this story many times as it was one of his favorites.

On another occasion someone brought a big rattlesnake to the barn and there must have been a dozen or so of us looking at and discussing the snake. Will was present and announced to the crowd that he wasn't afraid of but four kinds of snakes. Of course we expected him to name Alabama's four poisonous species of snakes. Someone made the mistake of asking Will what they were. Will said, "I'm only afraid of big ones, little ones, live ones and dead ones", which ended the discussion.

One morning at the barn, Will asked me if I would like to have a puppy. I told him I would and he took me to his home and showed me a litter of three puppies under his porch. "Pick one out and you can have him", he said. They were all males and I chose a lively brown one that looked like a ball of fur. I promptly named him "Brownie". He was part German Shepherd and who knows what else, but he was a wonderful addition to our family. Brownie grew up to be a fairly large dog and provided a lot of family protection as a watch dog. He loved all of us but he especially loved my little sister, Betsy. As a toddler, Betsy would sometimes sit down on Brownie pretty hard. Instead of biting or snapping at her, he would just lay there and whine.

Much like the movie "Ole Yeller", Brownie became a symbol of protection and "right of passage" into our yard day or night. My mother grew to appreciate this brown dog as a family icon and came to love him as much as we kids did. Once, he saved

Betsy from some stampeding horses that were exiting the hall of the barn by pinning her against the wall until they were gone. It was not uncommon to see Brownie leading Betsy around the place with her hand in his mouth. Everyone knew Brownie one way or another. A stranger would be wise not to approach the house unless someone contained the brown protector of the premises.

Times have changed since the days of the early 1940's. In those days, people would brag on their horse or mule or on how well their dog could fight. Brownie as a young dog constantly got whipped by his brothers, the two puppies Will kept, whose names were Hotshot and Jack. Hotshot and Jack would team up on Brownie and whip him bad often cutting his ears, face and neck. I would always try to prevent these fights between the brothers but sometimes they would catch Brownie at the barn or away from me. Brownie grew stronger though with age and was a fine specimen of a dog.

One day Brownie and I were going toward the house from the barn when I noticed him "bristle" as he looked up the road. There stood Hotshot and Jack looking for a fight. I tried to catch and hold Brownie but I was too late. The three dogs clashed at the top of the hill right in front of a cattle chute. Growling, biting and engaged in fierce combat, the three dogs rolled all over the road. Jack and Hotshot teamed up against Brownie. One would ride his back and one would chew on his head. They were whipping my dog pretty well until Brownie got Jack by the throat and slung him into about a 10" deep water pool that had collected in front of the cattle chute. Now all three dogs were fighting in the water. Brownie was holding Jack under the water and Hotshot was biting and riding Brownie's back. Jack began to drown but Brownie wouldn't turn him loose and when Jack did get loose he decided he had had enough so he ran up the hill and gave up the fight. Now it was one on one with Brownie and Hotshot. The rest of the fight didn't last long because Brownie sent both of them back across the field and under the porch of Will Bill's house. The fight was talked about for days and Brownie was famous in his own little sphere of influence. As far as I ever knew, they never fought again, nor did I ever see Hotshot or Jack far away from that porch.

Brownie, however, did have another significant fight a few months later. Larkin and I were in a mule wagon traveling south on Garth Road near the cemetery where a small bridge crossed a creek. Farther down near Four Mile Post Road lived a man by the name of Charlie Rice. Now Charlie Rice had a dog, bigger than Brownie, a black dog with some long gleaming white teeth. Everyone considered Charlie Rice's dog to be the biggest, meanest dog around. I never knew his name so we just referred to him as "Charlie Rice's dog". About the time we got to that creek, we saw Charlie Rice's dog standing square in the middle of the bridge. Brownie, who was trotting alongside the wagon, saw him also and ran ahead ready to fight. Both dogs stood on the bridge and eyed each other for a few moments. Larkin and I could only stop the wagon and watch the event.

The two dogs clashed on the bridge and bit, pushed and shoved all over the area. I thought this may be the end of my dog. Brown and black hair flew as the dogs bit and slobbered all over the place. Finally while locked in mortal combat, they fell off the bridge and luckily Charlie Rice's dog hit his head on the bridge abutment. This addled him and Brownie grabbed him by his thick neck and plunged his head under the water.

Much like ole Jack, Charlie Rice's dog lost interest as he began to suffocate and started looking for a way out of the fight. He finally got out and started running home to the protection of his home porch. Brownie followed for a while then came back to meet us. He was bleeding but proud. He stuck his chest out and proceeded the rest of the way much like the conquering hero he was. After this incident, there was no doubt Brownie was the king of what was by now being referred to less and less as "Garth's Hollow" and more and more as "Jones Valley".

Ray and Brownie – 1940's

Chapter Six – Kiddie Club and The Arrow

My social life during the war years was limited to church on Sunday and maybe to an occasional visitor during the week. I was too young to drive but I was having a good time raising bantams, turkeys, dogs, horses and anything else we could catch or think of. One social event of the week came when my mother would take us to the Kiddie Club on Saturday afternoon. The place for the Kiddie Club was the Lyric Theater in downtown Huntsville and its purpose was to watch movies. The movie was usually a western starring the likes of Roy Rogers, Tom Mix, Lash Larue, The Lone Ranger and others. Kids could get into the show for 10 cents; popcorn, candy and a coke only cost 15 cents, so for a quarter you would be fixed up. We would meet friends there, because everyone went to town on Saturday afternoon. Maybe you would even get to sit by a girl. We would take a bath, dress up and be looking pretty good for this weekly event.

One Saturday just after lunch, I took a bath and put on my best clothes which included my white "buck" shoes in anticipation of going to the Kiddie Club. I heard a knock at the door and there in the yard were five horses saddled and four of the boys from the barn. John K. spoke first:

"We's got to move dem cows out of the west pasture and we needs you to help us. We done saddled ole Kissie for you".

"John I'm all dressed up to go to town; I've even got my white bucks on so I can't get dirty moving cows", I replied.

"We'll open the gates and all that, all you'll have to do is sit on that horse and help us drive the cattle" he quickly replied.

"Well I don't know, how long do you think it will take?" I said.

"Aw, it won't take us but a half hour or so" one of the other guys said.

"OK, but I'll have to stay on the horse and out of the mud", I remarked.

"Den, let's go".

I told Mother I wouldn't be gone long and would be back in time to go to the Kiddie Club. With that we rode off at a trot to the west pasture. We were fortunate that when we found the cattle they were just leaving the shade and were headed for a big pond, about five acres in size, to get some water. All we had to do was drive them on the upper side of the pond and along a fencerow to an open gate to the fresh pasture. Everything went well and all the cattle, cows and calves, went in the gate as planned. This rarely happens as there is usually one or more of the cattle that has another idea.

We closed the gate and congratulated each other on our successful cattle drive. Only 12 or 15 minutes had expired, so my date with the Kiddie Club was safe. We started back and when we came to the pond, the boys decided to give the horses a drink. I said I couldn't do that but they assured me that I could hold my feet up and stay clean. So I let Kissie have his head and he waded out into about two feet of water to get a drink. I held my feet up and was doing fine until ole Kissie laid down in the water. It was a hot day and I guess he wanted to cool off. The boys also almost got wet falling off their horses laughing. After wading around in the mud and water for awhile, I got Kissie up and got back on him. My white bucks were no longer white and everything from my waist down was wet and muddy. It was almost worth it to hear those boys laugh all the way back to the barn. My mother even laughed when she saw me. After another bath and different clothes, I still made it to the Kiddie Club. The movie feature was Roy Rogers and in one scene he was riding Trigger across a stream. I thought about how lucky Roy was that he wasn't riding ole Kissie.

In addition to riding horses, raising dogs and chickens, and hunting and fishing, I loved to shoot my bow and arrow. I had an old conventional bow and some "practice point" arrows that us boys would love to shoot. We would make "hay bale" targets and have contest after contest. One thing we loved to do was to shoot straight up in the air and watch the arrow slowly lose the battle with gravity and plummet down and stick in the ground. I was doing this one day about 150 yards from the house in a big field when Betsy came out from the back porch. She was holding one of her kittens. She had three kittens which she named "My kitty", "Ray's kitty", and "Other kitty." I don't know which one she was holding but I got the idea to shoot an arrow close to her to scare both Betsy and the kitty. I made a beautiful shot; the only thing was I hit Betsy in her side. Both Betsy and the cat shrieked and even though it didn't hurt her other than a little scratch, she was hollering at the top of her lungs.

I didn't know when I started running toward her that mother had seen the whole episode from a window. In a few minutes,

Betsy was not hurting as bad as I was and I think she was rather enjoying mother's thrashing of her older sibling. Needless to say, I never shot arrows anywhere close to Betsy or anyone else with my bow ever again. Betsy's cats also survived the shooting and thrashing and "My kitty", Ray's kitty" and "Other kitty" all had little kittens and Betsy loved them all. The offspring of these three cats made for a lot of cats around the farm. In fact, kittens were our most productive crop but also the least profitable.

Chapter Seven – Mother's Management and the Train Trip

G.W. Jones & Sons Farm, as it was now called, survived under Betty's leadership during the war years. Crops were made, harvested and sold. Timber was marketed resulting in cleared land for future production. Also, the family was safe under the protection of our grandmother Lula May Bryant, Necy, Brownie and others. Betty kept a 45 caliber pistol in her bedroom with which she couldn't hit the side of a barn. Occasionally, however, she would go out in the yard and shoot up into some big maple trees just to make a statement to anyone who heard the shot. Larkin would help by announcing at the barn that he saw her cut a small limb out of the top of the tree. Of course a limb would be hit in the thick tree at every shot, but Larkin made it a good story.

Lula May was the lunchroom manager at East Clinton Street School, feeding about 500 kids a day, so she provided transportation for us back and forth to school. Lula May was a wonderful grandmother and person. She provided a real source of comfort and love for the Carl Jones family while he was away in the war. Having Lula May on the scene allowed Betty more time to manage the farm. Lula May was a godly woman who had a profound and permanent influence on all of us. Even today sometimes we get together and reminisce over our favorite "Lula May" stories.

Mortgage payments to Metropolitan Life Insurance Company were kept current with the soldier's pay and farm production profits. Betty was having to learn the art of raising agricultural revenue from row crops and livestock production. Sometimes she would write for advice from the boys as to which crops to plant. By the time she received an answer, most often the crop would have been planted and harvested. One crop that produced extra revenue was cane molasses. Sugar cane was fairly easy to grow and we had a cane press, pans, skimmers, etc. so Betty would put up several hundred gallons of molasses in tin cans each year. This went well until one year someone broke into the storage room and stole every can. After this she didn't put up any more molasses.

Personnel changed from time to time as sharecroppers and workers would come and go. Families by the name of Conley, Robinson, Everson, Hereford, Tanner, Rice, Lewis, Jones and others were some of the more significant ones. One family that came to us near the end of the war was the family of Man K. Everson. He had six children – John K., Doo-Boo, Bay Child, Maylee, Dufunny, and Pot. John K. had only a 2nd grade education but was one of the best cattle herdsmen we have produced. John had the knack of getting everything done, every day and could efficiently work as many men as you could give him. John worked for G. W. Jones & Sons Farm for over 40 years and when I preached his funeral, I felt as though I had lost a very close friend and family member.

Still another family that came to us in the mid-forties was the Harry Conley family. Stepson Luther Robinson was a wonderful man and responsible worker. Luther started working on the farm as a teenager and at the writing of this book, he is still working for us with over 60 years service. Most of Luther's service has been at the office; however, he was involved each June in the fescue grass seed harvest on the farm. Luther was and is a valuable worker at whatever the task. There is an old saying about working mules as a team.

54

The saying about a good mule is that "you can't hook him up wrong that he won't work". Many times, if not hooked up right or on a certain side, the mule wouldn't work. Luther Robinson couldn't be hooked up wrong because he would always work at his best, perfectly willing to pull 110% of the load, and with a good attitude.

With these and other faithful workers, the Carl Jones family endured the war and as a team, we all held down the "home front". I'm sure the strain and stress of it all was worse on Betty than anyone else. She only got to see Carl twice in the nearly five years of his war service. Once was in San Diego, California just before he and his unit were deployed to Alaska. Carl wanted Betty, Betsy and me to catch a train to San Diego and stay for about ten days before he shipped out. We boarded a train in Decatur, Alabama and headed west to see our family patriarch. Things went pretty well until we got to Houston, Texas. When we attempted to change trains, the conductor informed my mother that the train was full of troops, who had seating preference in war time. She argued and tried everything she could to get us on the train but to no avail. In despair at being trapped in the train station for possibly days, she sat down on a suitcase and began to cry. Betsy and I lingered nearby not knowing what to do. An Army Colonel came by and asked my mother what was wrong. Betty explained that she was en-route to see her husband in San Diego and it would be several days before another train was available. This Colonel immediately took charge and told my mother to get her things and her children and to take his arm and accompany him back to the train. He walked up to the conductor and demanded a place on the train for his "wife and children". The conductor informed him that he would be glad to but there was not one seat left on the train. The Colonel would not be deprived and asked if anyone was riding in the women's restroom? "No sir" the conductor answered. "Then put my family in there until El Paso and then some seats should be available." The conductor complied and I rode sitting on a trash can with my head against a sink all the way to El Paso while mother and Betsy lay on a blanket and luggage on the floor. Afterwards, we never saw the Colonel that helped us that day. He just faded away into the large crowd of soldiers in the train station. Sometimes in life we all encounter angels unaware, of which he surely must have been.

The rest of the trip was somewhat uneventful. I remember crossing the desert and since the train was not air conditioned, we had to wet handkerchiefs and cover our mouths because it was so dry. One other incident I remember was one morning several soldiers came by and asked my mother if they could take Betsy to the dining car for breakfast. She complied and Betsy, who was about three years old at the time, had her first date. The soldiers came back laughing and thanking my mother for letting her go. Betsy was a very pretty and cute three year old and she had surprised them by ordering "sauerkraut and buttermilk" for breakfast. You can take the girl out of the country but you can't take the country out of the girl.

We enjoyed the visit with my father in San Diego. We swam in the cold Pacific Ocean, went to Balboa Park Zoo and stayed until he had to leave for Alaska. Little did we know that it would be three years until we saw him again. The train trip home was easier since most of the troop travel was deployed toward the west. We had sleeping berths most of the way and looking back, it was a very courageous event on the part of my mother to take a seven year old boy and a three year old girl across the United States without guaranteed reservations on a train. When we did arrive back home at the farm, we were most thankful for the safe trip and every prayer uttered by those at the home front for our loved ones engaged in the war. The entire nation was totally dedicated toward winning WWII. Everyone who could bought war bonds. Women worked in shipyards and other defense industries to produce war material and they sold their jewelry to help finance the war effort. Families grew victory gardens, abandoned vacations and sporting events and prayed fervently asking God to protect their loved ones fighting this terrible war.

By the end of WWII in 1945, my mother had grown quite efficient at running the farm. She had the respect of the local farmers and businessmen who were not involved in the war. In addition, she was also known by the town females as a real genteel southern lady. An invitation to Betty's home was coveted by all and this continued to be the case until she died in 1990. She, more than anyone, was responsible for the survival and basic foundation of the farm operation which continues to this day. An era was coming to a close with the end of the war. The men were returning and the country and the farm would see many significant changes in the future.

Elizabeth B. Jones, "Miss. Betty", speaking at a cattlemen's field day.

Chapter Eight – World War II Ends

The city of Washington, D.C. was jubilant just as every other community and hamlet in the United States was on V-J day (victory in Japan). Betty, Lula May, Betsy and I had traveled to see my father by way of a 1939 Pontiac and gas ration stamps to Washington in August of 1945. V-E day (victory in Europe) had occurred on May 8, 1945 and my father was fortunate enough to be called back to Washington soon after the European Victory for a de-briefing of the war at the Pentagon.

With excited and light hearts, we had seized the opportunity to see my father and Betty drove almost non-stop from Huntsville to Washington. Upon arriving near the place where we were going to stay, the Frances Scott Key apartments, we saw my father walking down the street with some other servicemen. My mother stopped the car in the street and we had an unforgettable family reunion. A very vivid war memory of mine was seeing wives, sweethearts, parents and children saying goodbye or welcoming home their warriors. Even today I still "tear up" when I see our servicemen and women receiving affection from their families when returning from or leaving for military service. Such was the case in the middle of that Washington street. My father picked up Betsy to carry her a few yards to the apartment while mother retrieved the car. Betsy, who was about six years old at the time, had not seen her father much in her lifetime. She exclaimed excitedly when mother came back, "Mama this is my Daddy".

We stayed up most of that first night with each of us sharing homeland stories with our returning warrior. Mother shared the news of the farm, Lula May filled in with news from the community, Betsy with news about her kittens and me with news about Brownie, Kissie, Pooch, and the bantams. My father was gone to the Pentagon during the day but we made the most of our family time when he got back from work. We visited the Smithsonian, the Lincoln and Washington Memorials and other Washington sights with each of us having a grand time. The war with Japan was still raging even though the atom bombs had been dropped earlier on August 6th and August 9th. Little did we know that we would personally witness the V-J day celebration in our nation's capitol. The country was already thankful and relieved at the announcement of V-E day but it was extra special when President Truman announced that there was finally victory over Japan.

My father insisted that we all walk to Pennsylvania Avenue and the White House to join in the celebration. There was no need to attempt driving. The streets were filled with people, shoulder to shoulder, each celebrating in his own way. The policemen tried to keep people and traffic organized but were so excited themselves that they would lead a spontaneous cheer among various groups. More than once we saw a policeman stop directing traffic and kiss a pretty girl resulting in more traffic disruption. No one cared and all celebrated, similar to sports fans whose team had just won a big "come from behind sporting event". Several servicemen kissed my mother and wanted to hold Betsy. Some corporal came up and hugged my father who was a full colonel in uniform. My father just hugged him back. None of us had a camera but what I wouldn't give for a video of those scenes on V-J night in Washington, D.C.

Our visit lasted for a week or ten days and then we made the long trek back to Huntsville. All along the way and even when we got home, there were signs of jubilation and relief. The whole country was thankful that our land had been spared and that as a nation we had met the aggression of the Japanese and Germans successfully. Thankful that as a team of allies, with God's help, we had prevailed over the forces of evil. In subsequent years, military historians have analyzed almost every battle of WWII. In many of them, like "The Battle of Midway" and "The Battle of the Bulge", historians are at a loss as to why our side prevailed. In these and other battles, success should not have fallen our way except for Divine intervention. I believe the millions upon millions of prayers uttered by our side of the conflict were answered by the Almighty and lest we forget, this needs to be rehearsed in the ears of our children and subsequent generations. It was not insignificant to the war's success that President Franklin D. Roosevelt asked the entire nation to get down on their knees and pray for our men landing in Northern France on D-Day. Finally though, the war was over and the mothers who had flags with stars hanging in their windows depicting how many from that household were involved in the war, could take them down. My grandmother had a flag with six stars and one of those was gold. Gold stars delineated that that household had a member who had lost his life in the conflict. My cousin Nelson Jones was killed in Southern Germany a few weeks before V-E day. He is buried in Margraten Cemetery in Southern Holland near Maastricht, Holland. Libby and I have visited his grave twice. The cemetery is immaculately maintained just as are all of our national military cemeteries. Nelson is buried alongside 8,000 other soldiers who paid the supreme sacrifice for our country and its freedom; a grim reminder that freedom is not free.

It was about two months after our trip until my father finally got back to the farm. It took several days for him to visit with people who were interested in his service experience. Sharecroppers, businessmen, friends and family all made their way to the farm for a visit. After all, he had been gone for almost five years. He walked and rode all over the farm just trying to ascertain where we were and to think about the farm's potential. Even though Betty had held the farm together, the truth of the matter was that there was a lot of work to do to make it a viable enterprise. The row crops were being grown in small patches with very little continuity or rotation in the scheme of things. The few cattle we had were fairly good but the pastures were deplorable. That first year after the war in 1946, Carl and Ed planted some new pasture crops. Legumes such as crimson and button clover were planted but they quickly realized that the biggest need was a basic grass forage to go with the legumes. This began a nationwide search for an efficient pasture grass for the farm and little did we know that this search would eventually change the direction of our farm for the next half century.

V-J Day, the war finally ends in 1945.

Chapter Nine – The Changing of the Guard

The return of our servicemen dramatically changed our way of life. Our local population and family had settled into a relatively comfortable pattern. Work, school, church and other activities continued during the war but life was tempered with the longing that Hitler, Mussolini, Tojo and their evil conquest would soon be defeated. Now that those prayers had been answered and the men were coming home, life as we all knew it was changing. Huntsville had new leadership which had a positive effect on business, the elected officials and volunteer organizations like the Chamber of Commerce. The men came back with a "spring in their step" somewhere between being glad they lived through the conflict and wanting to provide a better life for their families. They were ready to make Huntsville and Madison grow. Almost immediately they became involved in the Chamber of Commerce and formed an Industrial Expansion Committee. They put up a valiant fight to get the L & N railroad to run their new North-South line through Huntsville only to see our sister city, Decatur, AL., win that prize. Another effort was to lobby and encourage our government to locate an experimental "wind tunnel" on Redstone Arsenal. This project was finally located in Tullahoma, Tennessee which was a great disappointment to Huntsville's leadership. Even though they were unsuccessful in these early attempts to foster more commerce and business, everyone was becoming aware that the leadership of our county was changing with the return of these valiant soldiers. There was a quiet feeling that under their leadership, things would get better for the sleepy little cotton town of Huntsville, Alabama, with its population of 15,000.

In a similar way, the farm was also feeling the impact of my father's return. A new "pecking order" was being established for the farm decisions. The men were now receiving orders from Carl and Ed as to their daily activities. Crop and livestock decisions were not being made by Betty, the lady in the big house, but by these two relatively new men on the scene. Even my dog Brownie was having to adjust. For years he had been the "protector" of the family and now this newcomer was giving him orders and was accepted by the family. Brownie lived several years after the war but I felt he was never truly comfortable with this change in his life.

The principal crop on the farm in the late 1940's was cotton. The sharecroppers had farmed this land with mules for decades. The bollweivel, Johnson grass and other inhibitors of the crop made cotton difficult to grow. There were no herbicides, pesticides, tractors or mechanical machinery to enable the cotton farmer as exists today. Gooseneck hoes and mule drawn plows were the only tools to fight weeds. Thanks to the fertilizer experiment station of TVA, we did have fertilizer that was marketed in cloth sacks. The most popular mix was 6-8-4 which represented six units or pounds of nitrogen, 8 units or pounds of phosphate and four units or pounds of potash. The farm women loved these cloth sacks which they used for cleaning and from which they made dish towels and clothing. It was not unusual to see some of the men or the kids wearing some of this cloth with the numerals 6-8-4 still visible on the garment.

Ed and Carl kept meticulous records on the cotton crop. Some sharecroppers were more efficient producers than others. No matter how hard a family worked to produce a good cotton crop, it seemed that insects and weeds were the limiting factors. In 1947, the farm produced a really good cotton crop, a bale to the acre. A bale of cotton is considered to be $500^{\pm}$ pounds of lint cotton. Lint cotton is what the farmer receives after his raw cotton has been sent through a gin. The gin would remove the seed and trash from about 700-800 pounds of field picked cotton and on the lint cotton the farmer received payment. I remember many times going on the cotton wagon to the gin at Lily Flagg just south of the farm on Whitesburg Drive. The wagons were mule drawn and sometimes we would have to wait all night in line for our turn at the gin. The cotton gin ran 24 hours a day and seven days a week during the fall cotton harvest to compliment the agricultural efforts of our area which predominantly evolved around the cotton crop.

One Saturday morning after the good 1947 crop had been harvested, Carl and Ed were talking about the crop at the breakfast table. The conversation went something like this:

"We made as good a crop as we can expect", Ed said.

"Yes we did. The workers and sharecroppers worked hard and the weather was favorable but we still lost money. How can we ever hope to be in the black financially because we'll not make this good of a crop each year?" Carl responded.

"I don't know" was Ed's reply. Then he added, "The only answer is to diversify with different crops and maybe a larger cattle herd".

"That's fine but it will require a larger capital outlay with fencing, livestock purchases, etc., not to mention the fact that we have yet to find a viable permanent pasture grass", Carl said.

"I'm going to seriously start looking. I've heard about a grass in Kentucky that sounds pretty good but I don't know much about it yet," Ed added.

"Why don't we right now admit that life is too short to raise cotton and from this day forward look for another prominent crop for our land?" Carl proposed.

This conversation over the breakfast table was a defining moment in the future of the farm. Ed was good to his promise and diligently searched for a permanent pasture grass. His search eventually led him to Pembroke, Kentucky where he saw what was called Kentucky 31 Fescue growing luxuriously green on a hillside in the winter. The grass had been discovered on a Mr. Sutter's farm in Menifee County Kentucky by Dr. E.N. Fergus of the University of Kentucky in 1931, hence the numerals 31, and he had fostered its planting throughout the state of Kentucky. The Kentucky legislature had wisely passed a statewide law prohibiting the sale of the grass seed in an effort to get Kentucky fully established with the grass before sales began. Ed and Carl, after further investigation and Kentucky farm visits, embraced the grass as our future basic pasture grass. The main problem was that by Kentucky State law, seed couldn't be purchased. Someone dubbed Kentucky 31 Fescue as "the wonder grass" early in its discovery period and this delineation is still used on bags of fescue grass seed sold today. Almost every farmer and cattle rancher in the Southeastern United States in the late 1940's was salivating for some of this "wonder grass".

The farm had by 1948 secured four AC-60 combines for the purpose of harvesting clover seed for sale. We had built a small grainery across the road from the 1915 vintage mule barn and could clean white, crimson and button clover. The grainery had one bulk seed hopper upstairs that had to be filled by carrying seed by hand up some very steep steps. The bulk product would then flow over a small A.T. Ferrell seed cleaner which would blow out the light material and screen separate the rest from the seed. The seed would fall on a concrete floor. It was then scooped and weighed by hand. The whole process was rudimentary at most.

Ed and Carl, being civil engineers by training, launched on a plan to drive the four combines and tractors to Pembroke, Kentucky and help several farmers harvest their crop in hopes of coming up with some Kentucky 31 Fescue seed. One morning the small "caravan" of tractors, combines and service trucks began the journey to Kentucky. Most of the men, all converted black sharecroppers, had never been out of Madison County, much less the State of Alabama. They stayed in Kentucky about two weeks, slept and ate in the field and all reported a grand adventure when they returned. The farmers let our men and combines harvest their fields and they would give us the straw, which wasn't illegal. We would then re-thrash the straw before asking local farmers to bale it for hay. The re-thrashing produced some seed which amounted to about 2,000 pounds total that was brought back home. Kentucky 31 Fescue at the time was selling for $5.00 per pound which would be an excellent price today but was enormous in the 1940's. From this small amount of seed, we planted about 80 acres and this small acreage grew into about 8,000 acres at the peak of our production in the 1970's. Like it or not, we were fast moving toward the cattle and fescue businesses that have sustained our farming operation for over 60 years.

Somehow, my father found a brand new army surplus D-7 bulldozer that had been packed before the end of the war for overseas shipment. We used this piece of heavy equipment to clear land, clean up hedgerows, drain and smooth the land and even prepare a seed bed for the new fescue fields. Some of the sharecroppers became proficient operators of the dozer. More and more we were adding tractors to our equipment and the mules and plows were become less and less. Some of our best sharecroppers refused to make the change from cotton to grass and cattle. One of our favorites, farm philosopher Will Bill Battle, decided to continue to sharecrop with mules for another farmer. Most stayed, however, and made the conversion to become cattlemen. The most notable were John K. Everson and Larkin Battle, both of which continued as mainstays for decades on the farm. As the fields were converted to grass, the names of the sharecroppers have remained attached to the individual working the cotton field at the time. Examples of these are shown on our farm maps today – John K. Field, Will Bill Field, James Leslie Field the Orsborne Cotton Patch etc. The long time era of cotton and row crops was coming to an end on our farm and most of us were delighted.

The changing of the guard and leadership on the farm affected all of us but in different ways. Mother was relieved in that the decisions were no longer hers. Lula May moved to town on Church Street to live with her mother, Big Mama and her sister Elizabeth whom we called "Auntie". Necy moved home also. Betsy continued to spend time with her kittens and dolls, busily dressing all of them in hats and clothes. Mandy still cooked for us and she, Larkin and Necy took most of their meals at our house. I was working on the farm in the summer, raising bantams, quail, turkeys, chickens and other animals. I had another Springer Spaniel by way of Pooch having her one and only puppy in life. I named the female Springer "Pudge" and she played an important role in continuing the blood line of duck retrievers that I raised for over 50 years. Kissie had gotten old and I was looking for another horse by 1949.

In 1949, I was 14 years old and in the eighth grade at Huntsville Junior High School which was the Old Wills Taylor school building that my father attended as a boy. My father had begun to take me duck and quail hunting and I was enjoying having him home. Necy and I raised a cotton crop together for a couple of years, I did the plowing and mule work and she did the chopping and we shared the picking duties in the fall. We made a little money but felt much like Carl and Ed in that "life was too short to raise cotton".

Carl and Betty had one post-war child on January 10, 1948. They named her Carolyn Tannahill. Somehow we all called her "Tiny Tutter". Some of the men at the office named her "Little Fescue" since our main effort was planting the "wonder grass" as fast as possible on the farm. Carolyn was a precious little girl and one that Betsy and I always thought was spoiled. I suppose every family sibling feels that way about the youngest child. Carolyn also loved and still loves kittens and would line them up along with her dolls in a closet to hold school each day. That little closet still bears the name of "the schoolroom" today. Carolyn graduated from the University of Alabama with a degree in education and eventually taught in the Huntsville Public School system. Her daughter Sarah, named for her grandmother Sarah Elizabeth (Betty), also earned a degree in education and taught in the Huntsville Public School system. Sarah was voted Huntsville's Most Outstanding Young Teacher in 1999. I often think about how fortunate those kittens and dolls were to receive such a good education.

We had three milk cows that fell my duty to milk on occasion and I decided early on that I would pass on being a dairy farmer. We killed hogs each fall when the weather was cool enough. We were all involved on hog killing day, especially mother. We cured hams and shoulders, made lard, soap, and cracklings using almost everything that the hog would produce. One of my jobs was to "smoke" the meat in the smokehouse. One day I got the bright idea that wet newspapers would make more smoke than the hickory wood I was suppose to use. I got a real tongue lashing when mother discovered what I was doing to make all that smoke, even though it did look impressive. My father was busy at the office and with community affairs in town, but I was delighted to have him home. The feeling of security and comfort of having a wonderful father like him as our leader is I'm sure unparalleled in the father-son relationship. The changing of the guard on the farm was very good for all of us but especially so for the son of Carl T. Jones.

In the early 1950's the farm changed from small fields and timber to pasture.
Below is the same view in 2009.

Chapter Ten – Pudge

The Jones family has always hunted together. Early on when my father was young, they primarily hunted quail. They had good bird dogs; our family usually had at least two. My uncle Edwin raised field trial English Setters but his interest was waning on bird hunting and moving toward duck hunting in the late 1940's. My Uncle Walter B. Jones was serving at the time as Alabama's State Geologist and had built a duck camp in the Mobile delta near Mobile, Alabama. My father was anxious to take me on a duck hunt so when Uncle Walter invited us to come for opening day in 1948, we were excited about going.

I was about 13 years old at the time and was training my first Springer Spaniel puppy whose name was Pudge. I wanted to take her with us on the trip so I asked if it would be okay. At first my father and Uncle Walter were reluctant but eventually I convinced them to let her go. Pudge was very obedient and I could get her to do just about anything. She would retrieve, kill gopher rats in the chicken house, run rabbits, be completely still while I was squirrel hunting and do a number of tricks on command.

Early one morning, my father, Pudge and I loaded our guns and equipment into a car for the trip to south Alabama and its delta. The trip took about seven to eight hours and Pudge curled up on the floor between my feet. We met Uncle Walter in Flomaton, Alabama where he had arranged for us to sleep in a friend's home as there were very few motels in those days. We greeted the man and had supper before I could get up the courage to ask him if Pudge could spend the night by my bed in the house. He looked kind of funny and I thought he was about to say no until he asked to see the dog. I got her out of the car and had her put on a performance for our host. I had her shake hands, speak, roll over, sit up, and stay while I walked off to retrieve my hat. The man then exclaimed that that was the smartest dog he had ever seen and that he would be glad for her to stay in the house for the night. When we went to bed, I put her on a little rug on the floor beside my side of the bed. My father told me later that every time he woke up she had her muzzle next to my face and as far as he knew she didn't sleep a wink.

The next morning after breakfast we traveled a long way on a rough road through some heavy timber to get to a landing called "Old Blakeley". Here we boarded a 20 foot long wooden skiff with a 25 hp motor for the trip to the duck camp. Our route took us by several bays and rivers until we reached Chuckfee Bay and Mallard Creek. The camp house was built on stilts in the marsh adjacent to some timber. It consisted of a long sleeping and cooking room on the left with a large deck connecting to the boathouse on the right. We unloaded our guns, equipment and supplies and then made up our bunks. Our next task was to get out five of the small flat bottom duck boats from the boat house. These would serve as our hunting boats for the trip. Each boat would be equipped with two oars and a "mud" paddle. The mud paddle was a thick heavy paddle that could wedge the boat along if the tide went out leaving the boat stranded on the silt. Things went well until we got the second boat out of its place in the storage shed. By now several of the 10-man hunting party had arrived and when they lifted the second boat, the deck came alive with big gopher rats. Immediately Pudge went into action, killing two of the big ones before anyone else could do anything. The men, particularly Uncle Walter, thought this dog was great and lifted the third boat to see what she would do. Pudge got two more rats out of that disturbance and before the five boats were all launched, she had 13 big gopher rats laid out on the deck. Pudge didn't know what a duck was because she had never seen one, but she knew what a gopher rat was and was glad to put on a performance. We hadn't even killed a duck and she was already the talk of the camp.

The next morning before daylight we loaded our hunting gear into our boats with each boat tying a bow line to the boat immediately in front of it. The lead boat had a 5 hp motor and as we passed a "pre-stuck" cane blind, the back boat would be untied so that boat could go to its blind. My father and I were in the second boat and after we floated our way into the blind we stuck the mud paddle in the silt at the stern to hold the boat in place. It was still dark and when shooting time arrived Uncle Walter would sound a fog horn, signaling to us that it was legal to shoot. All three of us, Carl, Ray and Pudge, were anxious. Pudge was whining as she began to see ducks flying overhead. I reminded my father that she had never even seen a duck so I didn't know how she would react. Finally shooting time came and my father told me to shoot a big mallard that had its wings set for the decoys we had placed earlier. Boom --- the duck splashed about 20 yards from the boat. Pudge looked at me and then to the flapping duck and I told her to go get him. She hit the water full force, got the duck in her mouth and retrieved it perfectly to the boat. I helped her in and she seemed real proud of herself. Our limit was eight between the two of us and Pudge retrieved every one as if she were a professional retriever rather than a country rat dog.

On the way back, several of the other hunters were bragging on Pudge. We rowed by Uncle Walter on the way back to the camp because we were through for the day. When we got even with him, he waved us over to his blind. He addressed me with a deep voice. "Son, I've got a wood duck down right here, see if Pudge can find him." "Fetch dead Pudge", I said as she jumped onto the reed island. She was gone quite a long time when we heard her barking like she was running a rabbit deep into the reeds. One hunter close by said that there were numerous alligators in those reeds and that I would be lucky to get my dog back. At this remark, I started calling her frantically. She quit barking and then .... silence. At least ten anxious minutes went by before she showed up with that wood duck in her mouth. Now she really was the hero of the camp and even the other dog owners who had experienced, professionally trained dogs were complimenting the "rat killer" from Madison County.

We stayed for two more days having a wonderful hunt but on the last morning the weather turned bad. A strong north wind greeted us as we made our way into our blind. We killed five or six ducks pretty quickly but we noticed that Pudge was having trouble swimming with the water being too shallow. I crippled a green winged teal and the duck was swimming downwind as fast as he could. Pudge was swimming and barking right behind him when suddenly I realized she was getting too far out in the bay to fight the strong wind and waves back to the boat. We left the blind and began to frantically row and mud paddle after her. The north wind was blowing the water out of the bay and several times we were almost stranded on the bay grass and silt. Pudge got the duck and between our effort and hers, we closed the distance and boated one tired retriever with a green winged teal in her mouth. We looked back at the blind and realized that the distance, wind and lack of water was such that we couldn't go back the way we came. My father made a command decision to turn toward Mallard Creek which was much farther but offered less wind and hopefully more water to float the boat. Rowing was out of the question since there was very little water. "Mud" paddling was our only means of propulsion so we used both the oars as well as the mud paddle. We made a little painful progress over the silt, traveling about 100 yards the first hour. We had begun to make a little more speed when I broke one of the oars. This was discouraging because when and if we finally made the deeper waters of Mallard Creek, we would have only one oar and the mud paddle to propel the boat. We were virtually by ourselves without another boat or person in sight. I guess I would have been very afraid except for the fact that my father was in the boat. As long as he was around I always had a feeling of confidence that everything would work out okay.

Not willing to take a chance on my breaking another oar, my father took the remaining one and told me to help all I could with the mud paddle. My arms felt very heavy after paddling an hour with that oversized mud paddle. We finally reached Mallard Creek and deeper water. Another hour and a half of paddling and we made it back to camp. Uncle Walter was getting ready to come search for us when we reached the dock. The temperature was dropping as we loaded our gear into the 20 foot boat for the trip back to Old Blakeley. I remember pulling a poncho over my head and shoulders which knocked some of the wind off Pudge and me. We loaded our gear in the car along with a cooler full of ducks about 3 P.M. and departed for Huntsville. Pudge and I slept all the way back home. How my father drove the 7+ hours by himself as tired as he must have been, I'll never know. Pudge had proven her metal; she had performed well and retrieved around 50 ducks over the three days, for us as well as some of the other hunters. To say that I was proud of this farm raised, rat killing, duck retriever of mine would be an understatement.

Pudge lived to be about 10 years old and retrieved numerous ducks, geese and other game over her lifetime. She had three litters of puppies. Her first litter was ten and her last two were eight each, some of which made fine hunting dogs. When Pudge died, I buried her in the yard of the big house on the farm. I erected a tombstone over her grave with simply "Pudge" on it. It's there today on the east side of the yard where my sister Carolyn now lives. Pudge started a blood line of Springer Spaniel hunting dogs that I kept going for over 50 years. I don't know how many dogs I trained during that time. Some of the more memorable ones were Smokie, Missy, Sissy, Prissy, Pudge II and others, all were females. One of the dogs in this line was the strongest, smartest, best dog I had since Pudge. Her name was Lady. She could run in a full gallop with an 11 pound goose in her mouth. Once on a hunt she retrieved 44 ducks and 22 geese in one morning. Lady was a dream dog, the kind that you get just once in a lifetime. I was fortunate enough to have several of these in the Pudge bloodline with which God had blessed me. Our family all grieved when Lady was killed by a truck; especially my son Raymond, Jr. who loved Lady very much. The day before she died she was happily pulling him around on a sled over a frozen pond. She was like a member of the family but gone with her was an era of wonderful canine companions that we grew to love.

There was an abundance of waterfowl in the 1950's.

Chapter Eleven – The Seed Business

The end of the 1940's decade found the farm effort busily engaged in establishing fescue pastures. The farm in Jones Valley was the first in our area to be converted from cotton to pasture, which was completely established by 1951. Ed and Carl were also searching for other lands during this time frame. There was a brisk demand for Certified KY 31 Fescue Seed and the two Jones brothers were racing to satisfy the demand. The existing little grainery was overwhelmed with the large amount of seed to which it was being exposed. Local seed cleaners could not be of much help because they were not calibrated for grass seed. The closest plant that could clean a large volume of grass seed was a seed plant in Cincinnati, Ohio. The farm had no mechanical "dryer" at the time which presented still another problem because the seed had to be dried to a storable moisture content before a cleaning plant could clean the seed.

Redstone Arsenal is just southwest of Huntsville and was used in WWII as a munitions arsenal. This arsenal was declared "surplus" (all 32,000 acres) after the war and was being offered for sale by the government in the late 1940's. Because of the nature of its mission during the war, the arsenal had several huge warehouse buildings. These buildings had good roofs, concrete floors and plenty of access for trucks. The arsenal leased one of the warehouses to the farm for the purpose of drying fescue seed. We would haul the seed in bushel size bags (about 35 pounds) from the field to the warehouse. Then we would pour the seed on the floor to a depth of about six inches and have men make parallel rows about one foot apart with their feet. The men would then walk 24 hours a day through the seed and stir the seed until they were dry (to less than 13% moisture). Several shifts of workers were required to accomplish this drying process in addition to the harvesting crew in the field. When the harvesting and drying process was complete, the seed was re-sacked by hand and hauled to the cleaning plant in Cincinnati, Ohio. At the time, we were "direct cutting" the fescue seed heads standing in the field. This was a very poor way to harvest the seed because we brought high moisture content seed into the house. All this, the long trip to Ohio and fighting a high moisture content in the seed brought about a drastic change in our method of seed harvest.

Ed and Carl decided that the only way to properly solve these harvest problems was to build our own drying and cleaning facility. In the winter of 1950 – 1951, we constructed a large grainery housing a six screen AT Ferrell cleaner with companion Hart Carter disc separators and six dryer bins. The building was constructed entirely with oak lumber, some of which came from the farm. The 8" concrete floor was poured on a rock base. The concrete was mixed by hand on the site. The rock base was from an old slave built rock fence that was hauled in from the mountain. Carl believed everything should be extra strong. Even though building materials were not as available in those days as they are now, the oak and extra concrete made up the difference. I was particularly pleased with the construction because as a 15 year old, my father took me out of school for a little over a week to work on the building. It's been 55 years now since that construction and I can still drive nails pretty well and I give credit to that hard oak lumber and the building of the big seed house.

The dryer bins had slatted oak floors about 30" off the concrete floor and were force charged with air and heat from a "wind tunnel" outside the building. Two high speed six blade 48" fans compressed the air into the tunnel which flowed under the slatted floor. A layer of full packed seed sacks from the field would be placed on this floor forcing the air through the sack as well as the seed. A one million BTU propane furnace heated the air at the fan intake so the air was dry and warm by the time it got to the seed. It was kind of a "Rube Goldberg" / "jackleg" solution but it worked perfectly and subsequently dried millions of pounds of fescue seed. We ran the dryer continuously during the seed harvest. Each morning the dried seed bags had to be unloaded from the dryer bin and stored in the back of the building until they could be subsequently cleaned weeks later. The unloading crew consisted of everybody, including the field harvesting men. These men would work hard but would laugh the entire time. Jokes, contests to see which team could unload faster than another, singing and poking fun at each other were the order of the day. The harvest crew consisted of about 40 men and they were all black converted sharecroppers. At this particular time in history, they were the happiest race of people on earth. They came to work laughing, laughed all day, sang songs and joked throughout the day and went home laughing. Dependent upon the weather, the fescue harvest would consume all of our men for the month of June and in some years beyond the 4th of July. The new building made for a much easier way of handling the fescue harvest.

Still another streamlined method of harvesting was the "windrower". This tractor drawn machine would cut the seed stems and windrow them into a swath about 18 inches wide.

The windrow would be left in the field for a few days to cure, allowing the seed to lose most of its moisture before combining. The combines had to be retrofitted so they could pick up the windrows as opposed to cutting the seed standing. AC-45 tractors pulled the AC-60 combines which required a "sack tying" man to ride the combine. This was a tough job and particularly when there was bountiful seed production. Used coffee bean sacks from Brazil were attached to catch the seed from the combine spout. Each had to be hand tied and stacked on the machine until the machine reached a designated "drop off" point. Sometimes there would be a long line of field tied fescue bags for the hauling crew to pick up and take to the dryer. Luther Robinson was one of our most productive haulers in this effort. I was about 17 years old in 1952 and Luther and I hauled almost 20,000 bags that summer. Sometimes we would load and unload these seed bags until 10 or 11 o'clock at night and start all over again the next morning at 6 A.M. The only relief any of us got for those 30 – 40 days of fescue seed harvest was when it rained. Luther Robinson is still working for G. W. Jones & Sons and is in his 63rd year of service at the writing of this book. In those early years, Larkin Battle and John K. Everson were the mechanics and coordinators of the combines and windrowers. Many farm hands, friends, associates and others were involved over the years in the fescue harvest. The annual production of fescue seed would eventually grow to one million pounds, all dried and processed through that farm seed plant that Carl and Ed built in the early 1950's.

The new seed processing plant under construction in the 1950's.

The seed processing plant today.

Chapter Twelve – Other Tracts of Land

The search for more land on which to plant fescue led us to the R.J. Lowe farm at Greenbrier, Alabama. We planted about 1,200 acres on that farm with the understanding that Mr. Lowe could graze the grass all year except from March through June or the end of the seed harvest. Things didn't go well because of over grazing and the travel distance to the Huntsville farm. After two years, we gave up this arrangement in favor of an 1,800 acre tract on Redstone Arsenal. This tract was much closer to the Huntsville Farm and served us for many years until we lost the lease in 1965. The "arsenal tract" produced good fescue seed and our cattle always wintered well at that location. John K. usually had the arsenal tract as his responsibility. There were no cell phones or two way radios in those days so John had to make decisions on his own once he left home. We ran about 500 mother cows on the lease and had our own hay storage, working pens, horses, etc. necessary to carry on the cattle operation. Many good stories came out of our arsenal lease experience. I'll relate one of the humorous ones.

One day Larkin and I were checking the cattle in one of the larger fields that was adjacent to Rideout Road. Arsenal traffic was pretty busy when this story happened in about the mid 1950's. At the time there were no coyote in our area, only red and gray foxes. All of a sudden as we came over a hill, there stood a coyote in full view. We were not supposed to have a gun on the arsenal but we always carried a shotgun under the seat of most of the farm trucks. Larkin was driving and quickly said "get the gun and shoot him". I retrieved the shotgun and put in a shell as we were now in full pursuit of the coyote across the rough pasture. I leaned out the window trying to get a bead on the bouncing and fleeting target. Finally, as we got near to Rideout Road, I fired off what I thought was a pretty good shot. Boom – the coyote rolled over with hair and dust covering the hood of the truck as Larkin slammed on the brakes to keep us from going into a little ditch. Momentarily the coyote staggered to his feet and ran straight into the traffic on Rideout Road. A car hit him and finished off the coyote and this caused quite a traffic jam. The military police showed up along with a photographer and other people interested in this strange, at least for Alabama at the time, road kill. Larkin and I slipped the gun back under the seat and pulled up to the fence in order to act like gawkers. The next morning the newspaper announced on the front page the strange phenomena of a man running over a coyote on Rideout Road. Today coyote sightings are a common occurrence, but at the time of this incidence I suppose to see one was a rare event. I'm sure glad it was the coyote that got recognized and not us.

My father made two other land purchases in the early 1950's that proved to be significant to the family. One was a run down cabin on Elk River near Rogersville, Alabama that required me, several farm hands and carpenters about three weeks work one summer to make it livable. This spot on the Elk River was our weekend family retreat. Water skiing, swimming, fishing and a lot of close family time has occurred at this special place for four generations now of the Carl Jones family. A new cabin was constructed in 2002 and as I write this part of "The Farm", it is 2009 and we are planning a 4th of July outing with the family in a few days.

Still another purchase was a small duck hunting tract in southeast Limestone County just off the County Line Road that separates Madison County from Limestone County. With farm personnel and equipment, we built a six acre "green tree reservoir" pond that could be filled just before the duck season began in December of each year. The tract bordered the Wheeler Wildlife Refuge on two sides. The refuge in those days provided a winter home for 75,000 or more ducks and about 60,000 Canada geese. My father had a well drilled and piping installed so we could flood the pond at an appropriate time for the duck migration. We planted pearl millet and corn in and around the pond every year and continue to do so to this day. Our kids, girls and boys, have enjoyed some fabulous hunting over the years as a result of this tract that we call "Swancott", named for a nearby crossroads on the map. The man who sold us this tract of land was Lawrence Devaney. Lawrence and his wife Ginny became good friends and remain so to this day. Lawrence was one of our hunting companions, a farmer by trade, and he planted and worked our land until he retired. Another couple that we became very close to in that same area was J.O. and Maude Marsh. Mr. Marsh looked after the hunting interest on this tract of land for many years until he retired. Mrs. Marsh was a wonderful person and a superb cook. We hunters and farm workers were the recipients of many good meals at her table. I can't begin to relate the warm spot in my heart that this tract generates when I reflect on its history. The memorable duck and goose hunts with family and friends and the wonderful association with good people like the Devaneys and Marshes are very special memories of mine.

Clearing land on the Swancott farm in the 1950's.

Chapter Thirteen – The Cattle Business

The decade of the 1950's saw many significant changes to life in Huntsville, Alabama. During Christmas of 1949, Dr. Werner von Braun chose the Huntsville Arsenal (subsequently named Redstone Arsenal) for the location of his German Rocket Design Team. Very wisely our government had convinced Dr. von Braun's team to side with the United States following WWII. Russia very badly wanted this rocket team and their intellect located on their soil. Primarily through the efforts of a General Lutie Toftoy, Dr. von Braun and his team were successfully encouraged to choose America for their future.

Initially the team and their families were re-located to White Sands, New Mexico. Almost immediately, the German families complained about the climatic conditions in this desert southwest area of our country. Frustrated with this unrest which could eventually restrict future rocket development, Toftoy convinced the Department of Defense to offer Dr. Von Braun a choice of six U.S. arsenals that had been declared surplus. Huntsville Arsenal was one of those six surplus arsenals and was chosen by the German team because the mountains and scenery most closely resembled the Bavarian area in Southern Germany. Almost unknowingly, the selection of the Huntsville Arsenal by the German Rocket Team forever changed Huntsville and this area of Alabama. This little sleepy cotton town with its population of 16,437 in 1950 would never be the same. Rural as well as city people were quickly affected by the move of these German families to our area. When the Germans first came, there were some ill feelings toward them as many remembered the atrocities of the war.

Gradually these feelings faded and these transplanted members of the von Braun rocket team became some of Huntsville's finest citizens. Huntsville quickly acquired the name "Rocket City". Blasts from the rocket firing test stands could be heard by most of the county and were a daily occurrence. Little did anyone realize that this area of our state was entering the space age which, at the writing of this book, has lasted for almost 60 years.

The farm, like everything else, was also affected by von Braun's decision. Some effects were good and some not so good. One good thing was that the farm's arsenal lease was solidified rather than being sold as army surplus. This was a wonderful tract of land and served us well until 1965. Both the seed and cattle production fared well on the arsenal lease. Traffic increased with the influx of more people coming to support the rocket endeavor at Redstone. There were some labor problems as some men left farm work for employment elsewhere. Even with all this new activity at the arsenal, the farm never wavered from the goal of expanding its seed and cattle business. Our fescue pasture acreage, both owned and leased, was about 2,000 acres by this time and hence there was a real need for livestock to graze these pastures.

Ed made a trip to Texas in 1950 to search for some female cattle to satisfy the need for livestock. Up until that time, we had relied on locally purchased cattle whose bloodlines were not very consistent. Ed settled on purchasing 400 head of Hereford heifers and it was an exciting day when we received his telegram with the message that they would be delivered in late June of 1950. Knowing they were coming caused a flurry of work. Fencing had to be secure, water and mineral points erected, extra hay had to be harvested and the heifer transportation to the farm arranged. Cattle were shipped long distance via train cars in those days because the trucks were small and the roads not very good. On arrival day at the Huntsville train depot, the farm had engaged six trucks, all single axel 20 feet long, to haul the calves from the train station to the farm. This went pretty fast because the haul distance was only about eight miles to the farm. I was only 16 at the time but I was driving one of the trucks back and forth to the train station. We unloaded in the big lot around the 1915 vintage horse barn until the cattle could settle down, eat and learn to drink our water which is quite different from Texas water. The cattle made the trip well and were good quality calves. Each calf had a brand of 6666 burned high on their rib cage. This reflected the ownership of the ranch from which we made the purchase of the heifers. The Four-Six Ranch was a famous old ranch in Texas breeding Hereford cattle and is still operating today. Supposedly the ranch was acquired in a poker game with a four of a kind hand of four sixes. In any event, it was a large ranch and bred good cattle at the time. These Four-Six heifers formed the base of our cattle herd which remained pure Hereford until we started cross breeding in the 1980's.

Even though the hauling of the calves from the train station to the farm went well, the process took all night. Most of the time was consumed with the railroad personnel re-positioning the eight cars hauling the cattle. About midnight during one of these delays, we truck drivers were visiting with each other around one of the cattle chutes when the subject of horses came up. One of the drivers, Mr. Gordon Darnell, asked me if I was looking for a horse. When I told him I was, he said "We've got time, let's go see a real pretty four year old mare

I've got for sale". I agreed and we got into a car and went to his house in east Huntsville where he had a barn and a training tract. Mr. Darnell whistled and this snow white beautiful mare ran to him and put her head in the bridle. He put a saddle on her and I rode her around the small tract three or four times and fell in love with this white mare. When the cattle were finally hauled about 3:00 a.m., I couldn't go to sleep for thinking about this mare whose registered name was "Cotton Lady". As promised, I called him the next day after consulting with my father. I had saved $125.00 working on the farm that summer but Mr. Darnell wanted $150.00 for the horse. At my father's suggestion, I offered $125.00 as a final offer and he took it. I picked her up that afternoon and was the proud owner of a registered Tennessee Walking Horse sired by the then well known "White Maury Boy" from Maury County, Tennessee. Quarter horses made the best cattle working horses but they were not widely used in the south in the 1950's, so we made do with horses of other breeds. I taught Cotton Lady to rope, open gates, and move cattle plus several tricks and we were a good fit for each other. She seemed to enjoy our being together as much as I did. We became the main ones to rope sick cattle or ones we needed to move to other places on the farm. One day we roped an 800 pound heifer that was in the wrong pasture. This was very foolish on my part because Lady only weighed 1,050 pounds. We fought this heifer all over the pasture with Lady putting all she had into holding this wild bovine. Of course once a calf is roped you can't just get the rope off the calf real easy. I could tell Lady was getting real tired so I found a strong fence post and tied off the rope to the post and loosed it from the saddle to give her some rest. It was hot and Lady was trembling so I left the heifer tied to the post, loosened Lady's saddle girt and led her back to the barn. When I got to the barn I put Lady in a cool stall and put the saddle on one of the two mules we still had on hand. The mule's name was "Laura" and she brought that 800 pound heifer to the barn like a puppy on a leash. Lady and that post had made a believer out of the heifer but I never roped one that big again.

On another occasion I decided to give Lady a bath while I was waiting on a farrier to come put some shoes on her. She was so clean and pretty that I decided to lead her up to the house to show her off to my mother. I didn't know it at the time but Mama was having an afternoon "tea party" for some of the ladies from town. I went around to the back of the house, opened the back porch screen door and called to my mother. When she didn't answer, I led Lady on to the back porch. The back porch had a concrete floor and Lady was barefooted so I opened the door to the house and leaned in calling mother to come look at my horse. I could hear women talking, the rattling of dishes and realized that a party was going on. I was almost in the house and Lady was nudging me to move on so I decided I'd show her off to the party. I led her down the hall and into the dining room. The scene in the main dining room was about ten ladies having tea and crumpets while laughing and talking. I led Lady through the dining room without saying a word and on down the front hall and out the front door. Several ladies gasped and one lady dropped her biscuit as the room went deathly silent. By the time Lady and I reached the front gate, I could hear some excited exclamations coming from the party. It didn't take long for my mother to find me after the party broke up. After she gave me "what for", I made her laugh by saying that "this will all be funny tomorrow". This story, along with other humorous farm stories, has been told hundreds of times over the years by our family.

Lady was a wonderful horse until her death in the late 1950's. Once I took her to Harlinsdale Farms in Franklin, Tennessee to be bred to the Tennessee Walking Horse Champion "Midnight Sun". She stayed at Harlinsdale for about three weeks before I got a call that she was ready to be picked up. I was regularly hauling fescue seed to Nashville at the time on a short trailer truck and on one of my return trips, I stopped by to get my horse. The farm manager at Harlinsdale said he would be glad to let her go but no one had been able to catch her for days. She was in a pretty big pasture and evidently several men had been unsuccessful in getting her to the barn. I asked him to give me a halter and drive me out to the pasture. He replied that he would do it but "nobody can catch that fool". When we found her she was standing by a water trough with some other mares. I got out of the car, spoke to her once and she put her head in the halter. I jumped up on her back and rode her to the barn bareback and with no bridle. When I arrived at the barn several of the Harlinsdale workers were scratching their heads in amazement that this "wild" Alabama white mare was so calm. She jumped right up into the truck and seemed glad to be going home.

A few weeks before she was to deliver a colt that next spring, she broke into the chicken feed house and ate too much chicken mash and died. A strange thing about horses is that they will overeat and get real sick and sometimes die, whereas a mule will not. I often wonder what kind of a colt she would have had from that jet black stallion. In the coming years, we had many wonderful horses. The cattle business and horses were almost inseparable, especially in those days. In the

1950's, all of our cow work was done with horses, however, today we use trucks, four-wheelers and driving fence wings. Today's method of driving cattle works, but it is not as meaningful to the mounted cowboy. I'm currently riding a quarter horse named "Doc" who is the best trained "cow horse" I've ever owned, but Lady always looms in my mind as the one who was an extra special part of my life in the cattle business.

Cattle grazing in the spring of 1950.
Note that the small white house is located next to the present Jones
Valley Elementary School. The same view in 2009 can be seen below.

Chapter Fourteen – The Early 1950's

The seed and cattle business, much like Huntsville, was expanding in the early 1950's. By 1955, the seed production from the farm and its homemade seed processing plant was about 500,000 pounds annually. The cattle business was also expanding. The Four-Six cattle were in peak production and the farm was becoming known in agricultural circles as a leader in the cattle business. We purchased local bulls for the farm mostly in our area of Alabama and Tennessee. This proved to not be a good decision because we infused some undesirable traits in the cow herd. In the late 1950's and early 1960's, this was corrected when we started purchasing Hereford bulls from Texas. The farm was regularly hosting field days for farm and agricultural groups interested in the cattle business. I was in high school at Huntsville High School from 1949 – 1953 and the farm and its seed plant were also the site of numerous square dances, barbecues and other functions. It was not uncommon to have several functions a month at the farm from the growing Huntsville community.

Huntsville was expanding rapidly with an influx of people from all over the nation. Good paying jobs related to the missile business was the draw and with more people came the need for housing and infrastructure. Our farm neighbors to the north, south and west began to develop their land into subdivisions and away from agriculture. The farm, however, continued to pursue expanding the seed and cattle business. By the mid 1950's, several of the 26 sharecropper families had moved away and there were only about ten of the houses that were still occupied. Families in these houses either worked on the farm or at the office of G. W. Jones & Sons in town. We were still baling small square bales of hay in those days so we used the vacant homes to store hay for the cattle business.

By that time some of the work force on the farm had also changed. Some of the new names were Buddy Tanner, Fred Lewis, Cleve Wesley, Johnnie Hereford, Robert Ayers, and our new overall farm manager U.G. Roberts. U. G. Roberts was a first sergeant for my father in Alaska and upon his return from WWII, he served for almost 50 years as the farm manager, with particular emphasis on the seed business.

One rather funny story I remember that happened about this time was my driving to town in our 1949 two door Chevrolet one day. Robert Ayers wife, Sylvester, had walked out to the still graveled Garth Road and was waving me down with her purse. "Good morning, Sylvester" I said as I stopped. "Good Morning, Mr. Ray. Wonder if'n I could trouble you for a ride to town?" She inquired. "Sure thing Sylvester. Get right in" I said as I opened the door and leaned up so she could squeeze into the back seat. I slammed the door as she sat down, put the car in gear shifting into a higher gear as we gained speed. I was just starting to begin a conversation with her when she tapped me on the shoulder and said "Mr. Ray, I wonder if you would be kind enough to open the door and let me get my hand out?" Startled, I opened the door and after inspecting her hand we decided that she would be okay. I never forgot this incident and still sometimes think about it when I hear a car door slam.

Mandy continued to cook at our house having done so for almost 15 years by the mid 1950's. Larkin, John K., Luther and others were continuing to work hard in the seed and cattle business. We calved in the fall (Oct. – Nov.) and harvested, dried and cleaned seed in the spring and summer months. The work was physically hard but we all seemed to thrive on it. The farm labor force was steady and mostly compatible. We did have one man, Roy Lee Ford, who was a bad apple in the bunch. Roy Lee was always bragging about how he was going to "whup" someone and was always in a fuss or fight. One night about 2:00 A.M., we got a knock on the door and my father and I reached the front door about the same time. There stood Otha Lee, John K's son, with a concerned look on his face. He said "John K. said to call the police because he done kilt Roy Lee". My father said "Are you sure he killed him?" "Yesser, he done shot him in the head". With this we called the police, got dressed and went up the road to John K.'s house. The police arrived shortly in two police cars along with an ambulance with lights flashing. One of the officers asked John what happened. John said, "I was asleep in the bed and this smartass boy came in and grabbed me and put a knife to my neck and told me he was going to kill me." "What did you do?" inquired the officer. John didn't hesitate, "I pushed him off'en me and retch up on the mantle and got my pistol and shot him". "Did you mean to kill him?" questioned the officer. "Yes sir, I sho'did" was John's answer. The questioning went on between the officer and John: "Where was the victim when you shot him?" "On the floor". "Where were you when you were doing the shooting?" "About two feet right above his head". "How many times did you shoot him?" "I think there was nine shells in dar". "You know I'll have to take you in to the jail overnight because of your intent to kill this man." "Yes sir". At this point my father spoke up and said to the officer. "Now wait a minute, John is a good man. He has never been in any trouble and besides he was in his own house and bed when this man threatened him". "I know Mr. Carl" the officer said "but there's been a murder with an intent to kill and I'll have to take him to the jail". About that time one of the other officers rushed up to the ambulance personnel and announced that the victim was not dead and should be taken to the hospital immediately. With that the ambulance loaded Roy Lee up and with sirens wailing, made for the hospital.

The arresting officer returned to where John, my father and I were standing stating that even though the man was not dead, he would still have to take John to jail. John had confessed his intent to kill Roy Lee so the officer continued to write out his report. With the aid of the police car headlights and a flashlight he scribbled his report. "What was the name of the victim?" he inquired of John. "Roy Lee Ford", came the answer. "Roy Lee Ford. You mean that man in the ambulance was Roy Lee Ford?" the officer exclaimed with disdain. "Yes sir". The arresting officer than called a caucus with the other officers, breaking the news that the victim was the infamous town trouble maker Roy Lee Ford. He then wadded up the report and threw it on the ground with the comment, "I hope he dies". That was the last John heard from the law about the incident. Roy Lee was back at work the next day with nine band-aids on this forehead. John K.'s only comment to the rest of the men at the barn was that he would never again load that 22 cal. pistol with "shorts" because in the future he would load it with the more powerful "long rifle shells". Needless to say we had very few dull moments on the farm and as a young man I couldn't have enjoyed life more. Plenty of work, play, adventure and an array of interesting characters made my boyhood life a real blessing.

I graduated from Huntsville High School in 1953. During my high school days I worked hard on the farm and dated almost every girl in the school at one time or the other. I drove that 1949 Chevrolet everywhere and it was an excellent car. The car was white and we named her "the powder puff". When I entered college in the fall of 1953, I was still driving the powder puff. Several of my classmates worked for the farm during the seed harvest and today we still remain friends. Men like W. F. Sanders, Dean Ratliff, Bobby Moorman, Topper Birney, Bill Goodson, Henry Anderson, and Bobby Tanner are alive today and still live in Huntsville. Bill Goodson, Dr. Bill Goodson now, was our class president and I was the vice president and we had a wonderful and very successful class of 104 students that graduated from Huntsville High School in 1953, about 70 of which are alive today. Recently, two of our granddaughters have also graduated from Huntsville High School and they are the fifth generation to graduate from that same institution. The first generation was my Aunt Elizabeth Yarbrough in the class of 1908, my mother in the class of 1928, me in the class of 1953, Lisa, my daughter, in the class of 1980, May Criner, my daughter in the class of 1982, Raymond Jones, Jr., my son, in the class of 1988 and Elizabeth and Allison Yokley in the classes of 2006 and 2008, respectively. If all goes as planned, the remaining grandchildren should graduate from Huntsville High School in the coming years. Needless to say, HHS has played a big part in all of our lives growing up in this small southern town.

Hereford cattle grazing in Jones Valley.
Notice the sharecropper's houses remaining from the "cotton days".

Chapter Fifteen – City and Farm Growth

Huntsville was growing so fast in the 1950's that the town was adding the equivalent of a new schoolroom per week. It was not uncommon to see 30 – 50 new subdivisions come before the planning commission of the City of Huntsville each month. Huntsville was the point of destination for the missile and high tech world. New people were coming to town daily from almost every state. Many of them had predisposed opinions about this backward state called Alabama and this one-horse town called Huntsville. Yankees were the most opinionated and were often unkind with their remarks. I remember one story about a particular Yankee that circulated around the courthouse square during that time. The story happened at the Central Café in downtown Huntsville one morning as one of the newly arrived Yankees came in to order breakfast. "What will you have this morning, sir?" asked the waiter. "What do you have on the menu that's good?" the customer asked. "It's all good". "I would expect you to say that dummy" the Yankee replied. "I simply want to know what is the best, your specialty if that's not too difficult for you". "Well", the waiter replied "We have some mighty good beef tongue". "Beef tongue! You mean the kind that comes out of an animal's mouth?" the Yankee customer replied. "Yes sir, that's the kind", said the waiter. Very indignant the customer declared "I'll have you know that I don't eat anything that comes out of an animal's mouth". "OK then, what will you have?" said the waiter. "I guess I'll just have a couple of eggs", responded the learned Yankee. Listening in to this dialog was the entire restaurant and the place exploded with laughter at the Yankee's last comment. I'm not sure the learned Yankee ever "got it" because he was on a different level.

Another rather humorous story that made its rounds in the community at that time was fostered by the Chamber of Commerce. The Chamber was working hard in those days to solidify keeping Redstone Arsenal in place. There were several cities and other locations that were politically trying to move the arsenal to their confines. One of the Chamber pleas to the community was to make life as pleasant as possible for the military personnel stationed at Redstone. "Invite them to dinner, open your homes, take them on outings, etc." was the Chamber's plea to Huntsville's citizens. One lady living in downtown Huntsville decided to do her part so she called an officer in charge at Redstone with this request. "I would like to invite some of the soldiers stationed on the post to Thanksgiving Dinner" she said to the officer. "Ma'am, I think that is perfectly wonderful", the obviously pleased commanding officer responded. She said, "I have room for seven soldiers at my table and I would like to have them eat the Thanksgiving meal here at noon on Thanksgiving Day". "Yes, mam," the officer responded as if anticipating a restriction as to who he might send. "Now there is only one restriction. Don't send any Jewish boys," she dictated. "No Jews, mam". "That's right, no Jewish boys". "That will be fine mam and I will have them there at your house at noon on Thanksgiving Day", the officer concluded. "Thank you officer". Thanksgiving day came and there was a knock at this lady's door. When she opened the door there stood seven fine looking soldiers dressed in their Class A uniforms and all the soldiers were black. The corporal in charge was the first to speak. "Good day mam, is this here (and he gave the address)?" "Yes it is," she responded. "Then we is supposed to have Thanksgiving dinner here," said the talking corporal. "There must be some mistake," she quickly said. Not to be denied the corporal repeated the address with the assurance that they were to have Thanksgiving dinner at that location. She responded again "there must be some mistake". The corporal said, "No mam, deh ain't been no mistake made because, Colonel Goldstein, he don't make no mistakes". Through the efforts of this well meaning woman and countless others, the arsenal remained intact and even today is still expanding. Redstone Arsenal is and has been one of the economic drivers of not just Huntsville, but the entire State of Alabama for the last 60 odd years.

The farm in the early to mid 1950's wasn't faring as well as Huntsville. The farm was struggling, particularly in the cattle business. The fescue was growing well and was producing seed and grazing. Seed and cattle prices were down and the cattle didn't seem to be gaining well. At first we blamed it on poor selection of sires, then on parasites but the truth of the matter, neither was totally at fault. It would be three decades until we found out that the root cause of the poor beef production lay with an unseen fungus in the grass. Absent this bit of information, Ed and Carl wanted to up-grade the livestock knowledge and management for the farm on a long range basis. I graduated from high school in 1953, loved the farm and everything about it so both Ed and Carl encouraged me to study agriculture in college. It took a lot of soul searching on their part because that meant going to the Alabama Polytechnic Institute in Auburn. The entire Jones family had attended the University of Alabama in Tuscaloosa and considered Auburn a bitter rival. Nevertheless, I entered

API in the fall of 1953, enrolled in Animal Husbandry and graduated four years later with a B.S. degree in that same curriculum. Meanwhile, Carl and Ed tried a number of things in an effort to make the farm more profitable. They changed the location of our sire purchases, only to make even more genetic mistakes. They worked on more parasite control and changes in calving dates as well as several pasture planting improvements. They introduced crimson clover and button clover into the existing stands of fescue, all with marginal success. While I was in college, they even went into the sheep business. A Mr. Mattingly came to Madison County and did a whale of a selling job on several farmers as to the profitable side of the sheep and wool business. Carl and Ed took the bait, hook, line and sinker. They purchased 500 western cross-bred ewes from somewhere out west and bred them to Suffolk and Dorset rams. They had a good lambing season averaging almost two lambs per ewe. Predators eliminated a lot of the lambs and their potential profit. Dogs, foxes but mostly skunks were the main culprits. Unlike most mothers, ewes will not do anything to defend their offspring from a predator, sometime she will stomp her front feet but nothing else. We sold the lambs at weaning time and clipped wool from the ewes in the spring. The farm employed a group of traveling Mexicans to clip the sheep and I got to help on weekends. I learned how to clip sheep and when I took sheep production at Auburn I "wowed" the professor because I was the only one in class that knew how to shear sheep. Sheep are not well suited for the southeastern United States because our soil conditions are too soft. Sheep are best suited for high country or rocky areas because their hooves grow about ½ inch a week. In soft textured country their hooves curl under and become infected. The only answer is to hand trim each hoof about every two weeks. My father gave me three men to help me and we trimmed sheep feet all one summer with a knife. To save you some arithmetic, 500 ewes have 2,000 feet and we began in June when school was out and were still trimming that fall. One of the happiest days of my life was when Carl and Ed decided, like the cotton business, that "life is too short to raise sheep" and loaded up every last one of them on a rail car headed west.

Sheep grazing in Jones Valley in the 1950's.

With the livestock business only marginally successful, the only answer for the farm was to concentrate on the growing seed business. The farm produced and packaged only Certified Kentucky Fescue Seed which gave us an advantage over producers of uncertified seed. Sales were brisk to points all over the southeastern part of the country. The farm became the feature story of several farm magazines and served as host to numerous bus tours of farmers from other areas. The home place (Huntsville Farm) and the arsenal were in good production but there was always the concern of losing the Arsenal lease so Carl and Ed began searching for another farm. Their faith never faltered that they could make G. W. Jones & Sons Farming operation a profitable enterprise.

Enrolling me in Auburn was part of the long range plan for the future livestock segment of the business. During my high school years I started raising Ring-necked Pheasant for sale to the Russell Erskine Hotel in town. By the time I entered Auburn, I had my own breeding pens, incubators, brood cages and a cleaning and marketing system worked out for my market. My one customer, the Russell Erskine, would buy all I could raise and pay $6 per brace for the pheasant. A brace of pheasant is a package with a hen and a cock back to back frozen ready to be delivered to the customer. I could gross $1,500.00 with expenses of about $500.00 per year. A year at Auburn in the 1950's cost about $900.00 total for the three quarters. I was fortunate enough to use this small enterprise to pay for all four years of college. Looking back, I now realize that my father allowed some of the farm help to feed and gather eggs for me while I was in school. In any event, the real value of my pheasant business was the business lessons I learned while dealing with the enterprise. Even though the pheasant business was small and elementary, it was very applicable to future business dealings that would serve the farm well in the days ahead.

County Agent Earl Halla and the author with a bag of certified KY31 fescue seed in the 1970's.

Farm manager, U.G. Roberts and a Herford bull in the 1960's.

HARVESTING KY. 31 FESCUE
SUMMER OF 1955

Combining Seed (Larkin Battle, Du Funny, Jo Willy)

Combining seed

Combining seed

**One of the 26 tenant houses that survived from the "share cropper" days.
This photo was taken from Garth Road looking East in the early 1950's.**

Carl T. Jones during construction of the cattle chute and squeeze.
This facility, constructed in 1948, is still in use today.

The cattle chute and squeeze today.

The current Betty Jones Garden Pump House and well.
It is located a few feet from the farm's original water well that was
"witched" in the early 1940's by "Miss Betty".

Betty Jones Garden – circa 2005

The "Environmental Steward Team" in 1996.
From left to right, Raymond Jones, Jr., Mike Patterson holding Caroline,
May Patterson, Ray and Libby Jones holding the Patterson twins, Will and
Bryant, Lisa and Mark Yokley with Elizabeth and Allison.

These are Hereford cattle grazing on the Huntsville Farm.
This photograph was taken looking SE in the 1950's and below in 2009.

Hereford cattle grazing on the Jackson County farm in the 1980's.

John K. Everson and young Raymond showing off a catch of carp.

Author and granchildren playing on large round hay bales which is always a popular pastime.

The Painted Bluff
Located on the border of the Guntersville Dam Farm in Marshall County, Alabama.

Elizabeth and Allison Yokley with the 2006 Champion Commercial Heifer

Ray and Raymond Jr. as GW Jones & Sons Farm hosts the American
Hereford Association Southeastern Field Day.

GW Jones & Sons Farm yearling steers.
These steers were featured on the cover of the Progressive Farmer
Magazine – August 1977.

Ray and Libby – September 4, 1960.

THE FARM
IN JONES VALLEY

PART II

Chapter Sixteen – Alabama Polytechnic Institute

In the fall of 1953, I entered the Alabama Polytechnic Institute, which was eventually named Auburn University. I enrolled in Animal Husbandry with the blessing of both Ed and Carl. Much effort had gone into the farm and both of the brothers felt it was good long range planning to have a family member studying agriculture. The summer before I entered Auburn, I joined the National Guard. I became a member of the 1169th Engineer Combat group and was attached to Headquarters Company located in Huntsville. The draft was still in effect even though the Korean War had wound down so the National Guard offered a little more stability than the volatile draft. Anyway, this was my father's old National Guard unit and even though I was at the lowest entry level as an enlisted man, I was proud to be a member of his old command. The National Guard duty was very compatible with college and farm work. Drills on Monday night, excused if you were in college, and a two week summer camp plus special duty if called upon. The only exception to this was if your unit was federalized for tasks affecting the country. The impact from my military service on my college experience was minimal while I was enrolled in Animal Husbandry at Auburn.

College life for me, as I'm sure it is for most students, was a real "growing up" experience. Even though I had been active in organizations at Huntsville High School, I was not well rounded in many areas. Academically I was weak in Math and English, failing the Auburn placement test caused me to take both subjects remedially for no credit. Socially I was a little shy having spent much of my life in the fields raising livestock. In high school I had been a member of Hi Y, De Molay, church youth groups and was Vice President of the senior class, but I still was not ready for college. I was fortunate that I joined Pi Kappa Alpha social fraternity and got to know about 100 men intimately. Their hometowns, vocations, and families became known to me and our friendships became real close and many of these men still remain best friends. I was active in the agricultural school at Auburn also, serving as Block and Bridle President my senior year and as a member of the livestock judging team my junior and senior years. The required agricultural courses I did very well in, even though I struggled with math and English. One of the most valuable courses I took was a sophomore class in English Composition taught by a Dr. Gosser. I learned more English and writing skills from Dr. Gosser than in all the rest of my English classes put together. I had some classic courses in agriculture also. Entomology with Dr. Guyton, Beef Cattle Production with Jim Orr, and Swine Production with Dr. Turney were some of the more outstanding ones. Representing Auburn as a member of the livestock judging team with Jim Orr as our coach was an education in itself. We traveled to most of the S.E.C. agricultural schools to compete in judging livestock. We traveled to competitive events in Chicago, Baltimore, Memphis, Atlanta, as well as several non-S.E.C. agricultural schools and we visited countless farms judging cattle, swine and sheep. As a team, we rated the agricultural schools we visited and at the time we felt the University of Tennessee and Mississippi State University were among the best but both were behind Auburn in quality. We all felt very comfortable with Auburn for our education as we strived to be effective in the field of agriculture.

One of the real values, both personally and professionally, of the Auburn experience came in the way of relationships. I made lifelong friends with several people that had a profound effect on me as well as the farm. Two of my best friends in life were fellow agricultural students that I met at Auburn. These two friends were Mose Tucker from Lafayette, Alabama and Roy Hereford from Faunsdale, Alabama. All of us served on Auburn's livestock judging team, eventually raised cattle and made several trips to Texas after college to purchase Hereford heifers and bulls. Mose is still raising and breeding some of Alabama's best cattle on his farm in Lafayette. Mose has been one of Alabama's most effective breeders over the years and is still doing most of the work and running his farm at the writing of this book. Roy Hereford became an outstanding auctioneer and operated a small cattle operation near Faunsdale in Marengo County, Alabama. Recently Roy was placed posthumously in the Alabama Livestock Hall of Fame for his significant contribution to Alabama's Livestock Industry. Roy is buried near his home in Faunsdale and when I attended his funeral as a pall-bearer, I felt as though I had lost one of the best friends I ever had. Still another personality from the Auburn experience that would have an impact on me as well as the farm was the judging team coach, Jim Orr. Sometime in my junior year I began taking Jim home with me on weekends. Jim fell in love with North Alabama, the farm and its cattle. Being a professor he saw several things we needed to do to improve our cattle operation. My father was impressed with his cattle knowledge, suggestions and organization. Eventually, my father engaged Jim as a consultant for the farm's cattle herd. Jim moved to Huntsville in the late 1950's and formed his own livestock consulting business and today still lives in Huntsville. One of the most needed improvements Jim Orr identified early on was the need for better bull selection to improve the farm's herd genetics. This need eventually led to several trips to Texas to purchase bulls and heifers for better seedstock in the herd.

I graduated from Auburn in June of 1957 and spent most of the rest of that year on Army active duty at Ft. Belvoir, Virginia. Being a First Lt. of an Engineer National Guard Unit in 1957 required that I attend a Basic Officer's Military Orientation Program (BOMOP) that fall. I was also required at the same time to attend the Army heavy equipment school at Belvoir. The school had us as officers actually operate and become familiar with the Army's heavy construction equipment. Graders, clam shells, dozers, back-hoes, trenchers, compaction equipment, trucks, etc. were part of the Army course study, all of which became valuable to me, the farm and our engineering business in later years. I returned home from Belvoir late that year and discovered that little or no provisions had been made to winter the cow herd. My Uncle Edwin passed away in June of 1956 and Carl was busy dealing with the loss of his brother and partner and settling his estate, so putting up hay for winter went unattended. The entire load of the farm and office and its many decisions suddenly fell on Carl. Ed's expertise in the cattle, engineering and abstracting businesses was to be missed as the firm, family and community mourned his passing. Carl was faced with the immediate decision of whether to proceed with purchasing a large farm in Jackson County on which Ed was working, the settlement of Ed's estate, and many other matters that would surface in the coming months. Additionally, the engineering, insurance, abstracting and the real estate areas of the business were beginning to grow, which further complicated the difficult days following Ed's death. Carl was to later say that those few years following the loss of Ed had to be the most difficult time in his life.

On February 18, 1958, North Alabama had an 18 inch snow and below zero Fahrenheit weather for about a week. The cattle were already weak from a tough winter and we lost several head during this winter storm. The area was paralyzed by this storm event. Roads were closed, utilities were out and few people had four wheel drive vehicles. Our tractors were ineffective so getting water and feed to the cattle became our most paramount need. In any severe cold spell like this one, the most important need for cattle is first water and then feed. They didn't teach this to me at Auburn and we mistakenly concentrated on feeding rather than water first and we lost several head as a result. I had a military driver's license at the time and borrowed an Army 6 x 6 truck to feed the cattle until the thaw came. In subsequent years, we always saw to it when a big freeze occurred that the cattle had water first and then feed. The cold weather cycle seemed to linger in the late 1950's and into the 1960's. We began to store extra hay, albeit the hay was still in small 50 lb. bales, to adequately handle the winters. The arsenal lease posed most of the problems during these cold spells because of the travel and lack of facilities on site. Frozen pipes, transportation of cotton seed meal pellets for feed, hay hauling from storage and removing ice from the water troughs would take most of the day for several men.

A couple of interesting stories came from these bad winter storms. One January day I was driving the borrowed Army 6 x 6 truck in the snow with a couple of boys from the farm delivering hay to the pasture when we came upon a small calf frozen standing up. The calf had evidently leaned upon a small bush to brace himself against the snow and wind and had frozen to death in an upright position. The black boys that were with me thought the calf was a "haint". This was black culture with reference to something that was supernatural or just real spooky. I said, "We better get this dead calf loaded up and hauled out of the pasture". The boys quickly responded "us ain't getting out of this truck cause that thing ain't no calf, it's a haint". So it became my duty to load the calf.

On another occasion, we were loading hay out of the old two story "Wind Jones" house one winter day when a convoy of about ten Army vehicles sped into the yard. This fine old house was part of the acquisition of the Wind Jones family farm acquired by the arsenal prior to WWII. One of the officers was a three star general with a swagger stick who was very much in command. The general walked around the house then approached me and said "Son, is this your hay in this house?" "Yes sir", I replied. "How soon can you remove it?" he queried. "We can have it out shortly General, we only have a few bales left" I responded in my best military fashion. Immediately the general wheeled around and barked at one of the lieutenants "This house will do, have it moved to the new location on the south arsenal (about 8 miles) and I'll expect to have a party in it tomorrow night". The astonished young officer replied "Did you say tomorrow night sir?" "That's what I said and I don't want to repeat myself again, any other questions?" the general commanded of which there came a response of "Oh, no sir". Within an hour, the Corps of Engineers had a large number of trucks, cranes, other equipment and men busily working to move the house. We got our hay out and within two hours the house was slowly creeping down the road enroute to the new location. The house is still being used today by the Army on the southern part of the arsenal. The general was General John B. Medaris who was assigned to Redstone Arsenal specifically to whip the country's space effort into high gear. The nation was in the thick of the moon race and Medaris was the right man for the job. His decisive command and leadership played an important part with America's winning the moon race.

Fortunately, the 1960's saw a new innovation that did away with the small 50 lb. hay bales which were labor intensive. The Vemeer company developed and produced the first large round hay bale machine in the 1960's which has been copied by other manufacturers and is still in production today. The round hay bales of 1700 pounds (equivalent to 35 small square bales) can be stored outside thus eliminating a lot of labor and storage requirements. They were quickly embraced by the agricultural world and also solved a lot of problems for the farm's arsenal lease.

While I was graduating from Auburn and serving at Belvoir, my father proceeded with purchasing a large farm in Jackson County. Upon returning from "The Engineer School" at Belvoir, my father brought me into the family business of G. W. Jones & Sons with the primary responsibility of the farms. The newly purchased Jackson County farm presented the most immediate need to be established in fescue pastures. Roads, fences, corrals, barns and pastures were built and established and after a five year effort, the Jackson County farm became similar in appearance to the home place and the arsenal leased farms. These added acres of pasture increased the need to expand cattle numbers for the farming operation. Jim Orr, Roy Hereford, Mose Tucker and I traveled to Texas and purchased several loads of hereford heifers from Bryant Edwards, an excellent rancher and cattleman who lived near Henrietta, Texas. Mr. Edwards and his wife Dorothy became good friends and visited in our homes several times in subsequent years. Through Bryant Edwards, we met several other breeders and cattlemen that had an effect on our cattle herd. We annually purchased bulls from several ranchers in Texas over the years. The P.J. Pronger family in Pampa, the Flat Top Ranch at Walnut Springs, and the Pete Hudgins family in Sherman were just some of the ranching families we met. Most of our dealings, however, evolved around Bryant Edwards who, in addition to being a "cattleman's cattleman", was one of the finest and hard working men I ever met. The last contact I had about him was with Dorothy, his wife, who told me that he was on a horse out in the snow in 20° temperature getting up cattle because it was too cold for his men. At that time Bryant was in his 90's and like most ranchers of his day he thrived on hard work. I still think about him often and the camaraderie all of us had on the Alabama end of this Texas relationship.

The cattle herd was growing and the foundation of the herd was being established and would last for several decades. Gradually the farm was infusing better bloodlines into the herd and we began developing a record-keeping system on each female that would also increase the effectiveness of our cattle herd. Somehow our efforts were not as productive as we had hoped. In the months of May and October, our cattle continued not to look good and their production failed. It would be about twenty years before we would find the answer to this problem that had persisted since Carl and Ed first planted fescue in the late 1940's. During this period of time, our only answer for improved production was to concentrate on better bull purchases, individual record improvement on the cows and innovative seed raising techniques for the farm. We were making some progress even though it was not up to our expectations. It was our hope that future production would improve, especially with all the work that had gone into the farm over the last 20 years. Personally, my hope was that the decision to send me to Auburn to study agriculture would someday pay dividends for the farm.

Chapter Seventeen – The School Teacher

In the early 1960's our fescue seed production began to increase. Jackson County, the arsenal lease and the home place were at peak production. We fertilized with about 60 lbs of nitrogen, phosphorous and potash in the fall and 60 lbs of nitrogen in the spring. Lime was applied to keep the ph (acidic level) in the mid to low 6s. The herbicide 24D was sprayed for broadleaf weed control every spring along with the insecticide toxaphene for army worm control. During this time we had enhanced our harvesting equipment with nine windrowers pulled by "H" John Deere tractors that we ran 24 hours a day during harvest. To thrash the seed from the field dried windrows, we had eight Allis Chalmers combines pulled by WD 45 Allis Chalmers tractors. It took about 35 men working the month of June to harvest and store the raw seed into the grainery. The grainery and its seed cleaning equipment would be used immediately after harvest to clean and package in 50 lb sacks the 800,000 pound crop. The seed was then sold in truck load lots to customers in most of the southeastern states. U. G. Roberts who had been with the farm since returning from the war in 1946, was instrumental in overseeing the cleaning and marketing of the farm fescue seed crop until we ceased production in 1984.

One incident that I vividly remember during those days was the all night windrowing of the crop. We would take a big truck to the field and tie a tarpaulin to its side making a "lean to" tent for the men to rest under. Hay was spread as a "floor" for the tent and covered with another tarp. The men worked four hours and rested four hours and rotated throughout the 24 hour period. Along about 2 a.m. each night that I had the time, I would go to a high place in the field and listen to the tractors and windrowers humming along in unison. I shall never forget the sound of the men singing at the top of their voices over the sound of the machinery. I know this sound can't be reproduced today because, like many other memorable sounds and sights, it too is gone with the wind. I wish I had had the presence of mind to have recorded that wonderful melody of men and machinery working together in harmony harvesting the crops in the still night air.

I was 25 years old in 1960 and was enjoying life. I was doing what I was trained for, which was raising cattle for the farm. I was living at home, enjoying Mandy's cooking, and going to National Guard drills. I had joined the Huntsville Rotary Club, and had lots of friends in town and over the state. I was a confirmed bachelor; having dated most of the girls in town I had announced to my mother that I was single and happy. Life was good.

During the fescue seed harvest in June of 1959, I got home about mid day one Saturday because we had been rained out. I was greasy, dirty, tired and hungry. My mother met me at the door and told me to get cleaned up and dressed because we were going to a wedding. "Mama, I'm not going to any wedding". "Yes you are, because I met a cute little girl from Cedartown, GA that I want you to meet. She's a school teacher and is one of the bridesmaids in the wedding," she decisively let me know. "Mama, I don't want to meet any school teacher from Cedartown or anywhere else, so I'm not going to the wedding" I strongly announced. "Get your clothes on because we are going to that wedding" was her command which she cast in such terms that the discussion ended. I suppose I subscribed to the old saying that "if mama ain't happy, ain't nobody happy", so I dutifully got dressed and accompanied her to the wedding.

As usual Mama was right because that little school teacher was "drop dead" beautiful. I was near the last of the crowd to go through the receiving line when I met this bridesmaid / school teacher named Libby. She was thirsty so I got her a cup of punch and we retreated to some chairs in the corner of the Russell Erskine Hotel ballroom. We talked, drank more punch and when we looked up, we were the only ones left in the room. I always said that she planned it that way because she needed a ride to her sister's house, so I took her home. Libby was visiting with her sister, Fran Strickland, at the time and planned to move in with her and her husband Strick that fall and teach in the Huntsville City school system. We dated regularly that summer with my having to stand her up several times because of farm work. We were both members of the Church of Christ so we attended church services and functions that summer as we have continued to do for the 49 years since meeting at that fateful wedding. Libby came from a big singing family named Mercer with five children; three boys and two girls. The Mercer family, Joe and Mary Ellen, had followed their family tradition and continued raising the current generation of Mercers in Borden Springs, Alabama in Cleburne County. Their homestead was rural and at the time I met and courted Libby, the only member of her family to graduate from college, they had no telephone or even a paved road. Libby visited Huntsville and her sister Fran several times that summer and I visited her from National Guard Camp located at Fort McCelleland which was not far away in

Anniston, Alabama. She had a wonderful family and a family life that was very spiritual and faithful to the church. More and more I was liking this school teacher who was beautiful on both the outside and inside. I suppose the one incident that impressed me most about her during our courtship was one Sunday afternoon when I took her to see the Swancott Farm. We still had our church clothes on so the idea was to just ride around and look at the corn and other crops. The corn was just right for "roasting ears" so Libby wanted to see if we could gather a few ears to take home. It had been raining but soon she was barefooted, wading the mud and gathering several dozen ears of corn to take home. I thought to myself that this girl was different than the others and really knew what was important, at least to a farmer. I can still see her in her Sunday dress wading in the mud in her bare feet. Our courtship continued through the next spring until I decided enough was enough. I was either going to ask this school teacher to marry me or not. We had enjoyed the fall and winter dating each other. Football games, horseback riding, ice skating, hunting, etc. is how we spent our time and more and more I felt she was the "one". I made the decision in a little grove of trees alongside US 72 on my way to check on a coal mine the family had a interest in near Fabius, Alabama. I had a long talk with the Lord, seeking His advice, while sitting on a log and I bought the engagement ring the next day. I had a date with Libby that night and when I got to Fran's house to pick her up, they were talking long distance on the phone to Mr. and Mrs. Mercer. The phone was in the bedroom and all of us were sitting on the bed so I thought it would be a good time to give her the ring, after all it would save a call later. When I put the ring on her finger she jumped up and started squealing "let me talk to mother". I always said "she fell out of bed" when I gave her the ring. All of them were fighting for the phone and I was left without an answer. Now almost five decades later I assume the answer was yes.

Our families were all pleased with the impending wedding. Larkin, Mandy, our friends as well as the farm and office employees were also pleased. My father had long since fallen in love with Libby and had remarked that if I didn't hurry up and ask her to marry me then he would ask her himself. We were engaged from the spring of 1960 until the wedding on Sept. 4, 1960. We were the first couple married at the new Mayfair Church of Christ building on Whitesburg Drive amongst a large host of family and friends. Our wedding reception was in the same ballroom of the Russell Erskine Hotel where we met. We had a picture made in our wedding clothes in the same corner where we had talked the year before when she was a bridesmaid and I was a bachelor. On our honeymoon trip we drove to Washington, D.C., New York, Niagara Falls and returned via Mammoth Cave in Uno, KY. When we had checked into our motel room in Kentucky, I came to the stark conclusion that we didn't have enough money to pay our motel bill much less enough to eat and buy gas for the trip home. I had not comprehended at this time that married life was expensive. The school teacher came to the rescue when she recognized that the owner of the motel was a "Quality Inn" associate of her Uncle Clayton Mercer who owned a "Quality Inn" motel in Huntsville. This guy loaned us enough money to see us home. I returned the money with a note of thanks a few days later.

I carried her over the threshold of a small farm house that had been constructed after the war in the 1940's. The house was constructed out of ammunition crates, the only lumber available at the time, and consisted of four rooms: two bedrooms, a kitchen and a living room. We had made a few improvements in anticipation of our marriage but we couldn't have cared less, because we were happy just to be with each other anywhere. We didn't have any neighbors nearby except for one man and his wife who lived about a quarter of a mile away. The black man's name was Buddy Tanner and his wife Ozella. One night about 1:00 a.m. there was a knock at the door which scared Libby. I answered the door and there stood Ozella. She told me to call the ambulance because Buddy Tanner had shot himself. After calling the ambulance I inquired just how Buddy, who worked for the farm, came to shoot himself. "He was cleaning his ole pistol", she said. "Where did he shoot himself", I continued. Ozella always wiggled her shoulders before she spoke. "He shot himself onest in his arm and onest in his leg and onest in his other arm", she answered. A few days later Buddy was back at work and Ozella came by to thank me for calling the ambulance. I took the occasion to tell her that I had heard from the other men at the barn that she was the one who shot Buddy. To which she put her hands on her hips, wiggled her shoulders and said; "Hummph, you liven to hear anything over here in dis hollow. I ain't tried to shoot Buddy since fo we lived down to Mr. Hay's place". Such was the introduction for my bride to our closest neighbors.

The school teacher, however, was up for almost every incident and has been my best fan and encourager over the years as times on the farm and office became tough. This honeymoon house she decorated, planted flowers and made into a wonderful home. Our first daughter, Lisa, was born two years after we were married. In 1964, we moved into our current house which is directly across the farm from this cottage. Our second daughter, May, was born that same year and when we left this little honeymoon cottage for a much larger and better constructed house, the school teacher cried. We both loved that little farm house and our chosen lifestyle of working the land and raising our family. Needless to say, the most wonderful command I ever received was when my mother ordered me to get my clothes on so I could meet this school teacher. She is and has been the love of my life. Libby, more than anyone else, has been the stabilizing influence in my life and has had much to do with the survival of the farm and our family business, which is 123 years old at the writing of this book.

The School Teacher.

Chapter Eighteen – The Expansion Years

The early 1960's found the farm still expanding in size and seed production as well as the cattle herd. A new tract of land in Marshall County, Alabama was purchased in 1960. This farm was 2,800 acres in size of which about one half of the 1,500 tillable acres was river bottom land and one half was upland. The Tennessee River bordered the tract on its west side and it adjoined TVA's Guntersville Dam just downstream. This farm, which we named the Guntersville Dam Farm, was very fertile and we immediately began to clear the land, fence and plant fescue on this place eventually making it similar in appearance to both the Home Place and Jackson County farms. These farms were within about a 30 minute drive from each other and the entire farm labor force was used, particularly in the winter, to transform this new tract into a cattle farm. Usually with calving season over in October / November and the seed harvesting and cleaning duties also concluded, all farm personnel worked on building another grazing unit for the cattle operation. Both the Jackson County and Guntersville Dam farms took about 4 – 5 years to complete their establishment in grass. Some land had to be cleared, mostly cut with a chainsaw, and the vegetation picked up and burned. The land then was tilled with a heavy disc, drained with ditches and smoothed to an extent that was practical. The next step was to lime, fertilize and broadcast plant with 30 lbs. of Kentucky 31 Fescue and 3 lbs. of white clover seed per acre. This process of planting was in the fall and the grass was available for grazing the following May. In addition to planting grass, water and mineral points had to be established and built. Fencing, driving lanes and cattle working pens had to also be constructed. It was no small feat to transform an old historic row crop farm into a productive grazing unit for the farm's cattle operation.

The Guntersville Dam Farm continues today as our best producer of grass. Our seed production accelerated after this farm was fully established and the sod was mature. By 1965 this farm was in full production which was a real blessing because in 1966, we lost the arsenal lease and had to completely vacate our cattle, buildings, corrals, etc. from Redstone Arsenal. It took over a month to disassemble the fencing, water and mineral points, squeeze chutes, etc. and transfer most of it to the Guntersville Farm. We had all grown fond of this arsenal tract of land and we left with a lot of memories of the past 18 years. In less than two years, we were back to full production having absorbed the loss of the land into the Guntersville farm. Also, within two years of losing the arsenal lease, a tragic event would occur that would highlight the fact that losing the arsenal lease was a blessing and I think the influence of Divine intervention.

Seed production in the mid 60's was about 750,000 pounds annually and our cattle numbers were over 1,000 mother cows. More and more G. W. Jones & Sons Farm was being recognized by farm groups. Bus tours, field days, magazine articles, etc. were becoming more frequent and caused a drain on our top personnel who were "minding the store". My father was an excellent public speaker for these visiting groups as well as for the community. Most of the leadership of the farm at the time was involved in building the cattle and seed operation and welcoming visitors to the farm. More and more, however, we were finding farm labor very difficult to find. Many of our original black families were moving off the farm taking new non-farm jobs. Many of the fine young black men during this time moved to Detroit, Michigan to take on better paying manufacturing jobs. The result caused us to begin to search for labor saving methods for the 35 men required to harvest our seed. No longer could we physically move 25,000 bags by hand with our current work force. Our work force was changing and in some ways it was a change for the better.

Some of the most interesting new employees of that era were men like Clarence "Four Mile" Lanxter and James Leslie. Four Mile Lanxter was a hard working black man with a jolly spirit. Jim Orr found him in South Alabama and helped move Clarence and his wife Josephine to Huntsville. Clarence could shoe horses, rope calves and do with gusto any kind of farm work. He had no education, signed his name with an X and only had one real drawback, which was his wife, Josephine. Josephine was always in trouble, she was a big drinker and caused all manner of problems for the G. W. Jones & Sons farm community. Once after a fight with Four Mile she got after him with a chainsaw cutting him up pretty bad, but she was the one who landed up in the hospital. The accounting office of the hospital called me a few days later and asked if she lived on our place. I told them that she did live on our farm, at which time the accountant said she was ready to be released and we would have to pay her bill in order for her to be released. After much discussion, I informed the hospital accountant that we would not pay the bill. "You don't understand, Mr. Jones, we will not release her until this bill is paid", the accountant responded. "Well then just keep her," was my reply. "How's that?," she replied. I repeated, "You just keep her then", after which there was a long silence.

Within the hour, Josephine was back home. Clarence eventually got another wife named Linda but she was worse than Josephine. In spite of his wives, everyone, especially our children, liked Four Mile who always had a winning smile and a positive attitude about life.

Another real character that came to us about that time was a man named James Leslie. James Leslie weighed about 125 lbs. soaking wet but could out work any man twice his size. James lived on the farm with his wife Daisy. One of the fields on the farm today is still named the James Leslie Field as he was the last one to live in a house near that field. James was an alcoholic and later in life realized his mistake, became a member of AA and was one of their most prolific speakers in the North Alabama area. James was a real character in his drinking days. Once in the fall of the year, U. G. Roberts, our farm general manager, and I were returning from the Guntersville Farm late one afternoon when we noticed a fire in an abandoned house near the barn. Instantly we rushed to the house and found James Leslie, obviously drunk, with a roll of burning newspapers trying to light a butane heater. "James, what in the world are you doing?," I asked. "Hey Captain, Messer Ray, I's trying to light dis here heater so de pipes won't freeze tonight", James stammered. "James, there's no gas to that heater or water in the pipes and you might burn the house down so put that fire out. Anyhow James you've been drinking," I said. "Drinking?" he responded indignantly. "Yes, James you've been drinking and there's no need to argue otherwise," I answered. "Well, Messer Ray, I'm not gonna argue wif you cause I can see dat you got yo mind made up about dat. Anyhow de real reason I come down here was to see if'n you would loan me two dollars," James confided. "James, you know I don't loan any of you men money if drinking is involved," I responded. "You mean you won't loan ole' James two dollars when I done worked for you in de hot summer and de cold winter and you won't loan me even two dollars?", James inquired. "No James, not two dollars or two cents while you are drinking," I informed him. "Well OK den Messer Ray. I don't need you anyhow Captain. You see dis here, dis here is my own bank checkbook", James said as he pulled out a small booklet from his shirt pocket. "Well then James why don't you write your ownself a check for two dollars?," I quickly asked him. "I can't do dat Messer Ray, cause I don't write no small checks," he informed me. With the fire out and Mr. Roberts and me struggling to contain our laughter, we left James with his checkbook and made an exit to the barn. This story was told all over the valley and is still enjoyed by many even today.

There are many other stories that permeate the history of the farm in that expansion era. During those days I was really enjoying life. The employees, the challenges of building these farms and the cattle herd made the long hours working outside on the farm enjoyable rather than being a burden. I also had another dimension to my working life that came from being in business with my father. I had time in the winter to learn and perform land surveying duties through the engineering side of the business. Eventually, I registered myself as an Alabama Registered Land Surveyor, Alabama RLS #6946. This expertise came in handy over the years with farm land line questions that occasionally surfaced. Another winter learning experience that presented itself during the mid 1960's was developing a rather large coal mine near Fabius, Alabama in Jackson County. Our family owned a mineral rights tract with the Dr. Milton Fies family of Birmingham. The mine eventually developed into over one million tons per year production of steam coal which was sold to the Tennessee Valley Authority steam plant and nearby Widows Creek. Dr. Fies and I became real close friends as he was a wonderful man and a wonderful teacher. On several occasions I was able to use farm personnel to help in the off season at the mine. On one occasion I was running a property line for the mining effort on a small tract with a "put together" crew of black boys from the farm. My instrument man on that occasion was named Buddy Everson, brother to John K. Everson mentioned earlier, and we were measuring and cutting this particular land line pretty fast. I had my back to the front of the line looking back at Buddy who was lining the front rod while peering through the transit. All of a sudden Buddy's eyes got big and there was a look of shock on his face. The chainsaw near the front of the cut line went quiet and in its place I heard and saw an Indian in full dress doing a war dance between us and the front rod. Everything went stone quiet and the Indian danced off and out of sight on a little path to our left. Buddy was the first to speak. "What was dat?" he said. "I don't know Buddy but it looked like an Indian to me", I responded. "Dat's what it looked like to me too and I'm gone", Buddy replied as he and the rest of my crew left me in the woods with all the equipment. I knew they wouldn't go far because I had the keys to the truck. I retrieved most of the equipment and piled it next to the transit. I don't know why I did this other than to make it easier for the Indian if he came back. I went back and found my crew milling around the truck and they announced to me that they wanted to go back to Huntsville. I told them we would go home as soon as we finished our surveying job. Right then I encouraged them to eat lunch and then we would talk. During lunch they began to laugh and joke about the Indian and within an hour we were back running the land line. It turned out that the guy was a real Indian who lived nearby in the woods. His name was Collier and he was mentally a little "off". In the coming days, the boys visited with Collier at his "teepee" and enlarged the tales told back at the farm.

All of our enterprises were flourishing by the mid sixties under the excellent leadership of my father. Carl Jones had the knack to manage most anything from farming to engineering. His excellent leadership and foresight projected G. W. Jones and Sons to heights never imagined by the early founders. Huntsville was booming and Carl was the man of the hour. Twice he had served as president of the Huntsville Industrial Expansion Committee. The community had awarded him "The Distinguished Citizen Award" in 1965. The county and city leaders had such confidence in Carl that they would often send him to make a pitch to prospective industries on the community's behalf. Community leaders, politicians, business associates, and most everyone who knew "Mr. Carl," as he was affectionately called, wanted to have his opinion on things. One of the most significant professional feats Carl was called on to perform that would touch thousands of people, was the job of designing the Huntsville-Madison County Airport. Through Carl's guidance and under the direction of the Huntsville-Madison County Airport Authority, G. W. Jones & Sons took 3,000 acres of raw land west of Huntsville and designed an airport complex that has become known throughout the world. Boasting parallel, 8,000 foot runways one mile apart with accompanying taxiways, a hotel, a golf course, and an industrial complex, the airport has grown from its opening in 1967 to a passenger usage today of over 1.2 million landing on 10,000 feet of runways. The airport complex designed by Carl and the "office" force received the National Achievement Award from the American Consulting Engineers Council in 1968 as well as numerous other awards and recognition since that time.

During this time, the farm also matured and established itself because of the overall leadership of "Mr. Carl" who was just as comfortable performing as a farmer or as an engineer. He had a saying that said "there is no substitute for leadership". Little did I know how meaningful this saying would be for me in the near future.

Fescue seed grainery and seed cleaner.

Chapter Nineteen – Difficult Days

I suppose that the one "constant" in life is the fact that there will always be change. The ten year period from 1957 to 1967 had been very stable for me as a young man. I was doing things that were fulfilling from a professional standpoint. I had married a wonderful wife. I was establishing myself in the business community and all under the watchful eye of my father. I had a good "friendship type" relationship with all of our employees (almost 100 at that time) as I intertwined with them as Mr. Carl's son. Over the years, other sons in business with their father have described to me the son's position as special and unique in the business world. Looking back I can clearly see that this was a golden period for me and I'm glad I recognized it even at the time. With a man like Mr. Carl as a backup, problems were not problems at all but merely welcome challenges that needed to be solved.

On October 7, 1967, Carl Tannahill Jones passed away during the second quarter of the Alabama – Ole Miss football game in Birmingham, Alabama. At age 58 he was at the pinnacle of his career, and his family as well as the community mourned the loss of this outstanding leader. Sympathy was extended to the G. W. Jones & Sons family from all over the nation. Friends, family, employees, business associates and community leaders for weeks mourned the loss of its spokesman. The Huntsville Times Newspaper had an editorial several weeks following Mr. Carl's death that stated, "The death of Carl Jones seems to still hang heavy over the business community." The community honored him by renaming the Huntsville-Madison County Airport the "Carl T. Jones Field" shortly after his death. The end of an era had come to a close for G. W. Jones & Sons as well as for the community, the family, and those who knew Mr. Carl. During his life, he had served as a father, soldier, engineer, farmer, businessman, banker, leader of the G. W. Jones & Sons family, and one of Huntsville's most beloved citizens.

On October 8, 1967, the day after my father's death, I assumed a new and unfamiliar role in the family business of G. W. Jones & Sons. I went from being the son of the business patriarch to its leader. I was completely familiar with the farming side of the business but with the "pecking order" being changed, even some of the simple farm decisions became difficult for me. The engineering side of the business found me with a vague knowledge of everyday activities. My father had once confided to me that if anything happened to him that I should immediately sell the engineering business. I contemplated all of these things in the days following his death. At issue and in need of leadership were not only the farm and engineering businesses but a Fire and Casualty Insurance Agency, a Real Estate business, an abstracting business and a Lawyers Title Insurance Agency. I was unfamiliar with most of these businesses since my degree and most of my work effort had to this date been involved in agriculture. Very little peace of mind or sleep presented itself in the difficult days following Mr. Carl's death.

I plowed ahead with the farm and its decisions through our managers. I met with them each morning at the farm office at 6:30 a.m., either in person or via phone. We would review the events of the previous day and plans for the next day as well as the future. I found that a hard hurdle for me was not now being able to have a "hands on" approach to things. I also struggled with having to be satisfied with a lesser quality of work than I had been accustomed to. I changed some personnel assignments that helped cover the need for extra leadership. I designated a farm manager for each of the three farms, and a general manager and an equipment manager as the cadre of leadership for the farm. Leon Miller had come to work for G. W. Jones & Sons in 1959 and in addition to being manager of the Jackson County Farm, I named him as overall herdsman. U. G. Roberts remained as general manager, John K. Everson as manager of the Home Place and Sonny Pritchett as manager of the Guntersville Dam Farm. A typical day for me was an hour or two at the farm office, home to eat breakfast with Libby and then to the downtown office to wrestle with business matters of a world other than farming. I would respond to questions on the farm via telephone during the day and at night. I shunned night meetings as much as possible so Libby and I could have some time together with Lisa and May. My father's words, "sell the engineering business if anything happens to me", rang in my ears as I contemplated a long range plan for both the office and the farm. Several weeks after my father's death, I realized that our average tenure of office personnel was over 21 years. One woman had been there 50 years, two people had over 40 years service and quite frankly, I didn't have the fortitude to tell any of our faithful employees to go home. I didn't ask for the position I was in but since it was my "time at bat", I felt that I should at least take a swing at trying to preserve our family business. I assumed the leadership role for both the office and the farm while trying desperately to fill Mr. Carl's very large pair of shoes.

Even though awkward, most of our employees accepted me in the leadership position. The same seemed to be true for business associates all over the town. I was asked to take my father's place on the First National Bank Board, a position I held for the next 35 years. Quite a number of Huntsville's leaders took an interest in me and knew of my struggles in trying to run our businesses. Men like Louis Salmon, Henry Bragg, Beirne Spragins, Bob Lowry, Pat Richardson, Tom Thrasher, Charlie Shaver and a host of other successful businessmen were a real stabilizing influence on me at this difficult time. Within 15 months after Mr. Carl's death, I had buried nine members of the G. W. Jones & Sons family. Some were family blood members and some were employees. One died in my arms at the office, one drowned on the Jackson County Farm, one member of the survey crew was electrocuted and "Pick" Johnson, one of my mentors and G. W. Jones & Sons general office manager, died suddenly of a heart attack in 1969. It seemed that everything I tried to get going in the office or on the farm was curtailed by grieving and going to funerals. The last death in this line of departure was "Pick" Johnson who was my "right arm" and a wonderful office manager during this time. I began to doubt whether or not we could pull this little 80 year old family business back from the brink of all these losses. I was working 18 hours a day, sleeping very little and grieving over the situation with all these deaths. To make matters worse, two of our partners in the real estate business filed suit against the family for division. All these things made for a lot of pressure on me and at best we were just "treading water" at G. W. Jones & Sons.

I began not to feel well and at the insistence of Libby I went to our family physician, Dr. Ellis Sparks, who did a thorough physical on me. His tone of voice was very grave as he gave me his assessment of my health. "Ray, I project that you have about six months to live. You are killing yourself with the pressure of work you are pursuing. The choice is yours, of course, but if you want to live, your lifestyle must change." I left his office and rode around the farm for the rest of the day and prayed. For the next couple of days I did very little work, cancelled a few meetings, read the Bible, prayed some more and did a lot of soul searching. One night in the wee hours of the morning I completely turned over the business and my life to God. I was finally convinced that I couldn't do it alone, it was foolish to have even tried by myself and I needed God as my partner and guide.

For the first time in almost a year and a half, I slept comfortably for eight hours. I decided that I was going to make decisions as best I could while asking for God's help. I really meant that I turned it all over to Him. I started every day by reading the Bible and asking God's help for that day. Even though it's been almost 40 years since that night that I turned things over to Him, I still follow this same practice. Within a few days after this decision, I fired a man at the office who was stealing from us and I began to assert myself to the G. W. Jones & Sons organization. The employees on the farm and at the office noticed the change and seemed to welcome having someone in command, even if it wasn't Mr. Carl. The real one in command was God working through me. I began to make engineering and management decisions that were well above my level of education and training. Several employees were surprised when I was proven right on problems that presented themselves. I made engineering presentations to boards, agencies and customers with ease and with the Lord as my helper I surprised everyone, including myself. I was named Huntsville's "Young Man of the Year" in 1968 and was becoming more comfortable with my dual role of being an engineer/farmer. I had resigned my National Guard Army Commission by this time which gave me more time for the business and family. I really liked the army and its discipline and structure. I had attained the rank of major when I left the guard and had made a lot of friends that remain good friends today.

The farm personnel remained fairly constant during this time as we continued to raise cattle and fescue seed. Labor was becoming more difficult and I spent a good bit of thought as to what we could do to solve the problem. Office personnel on the other hand were changing. Some employees left because of the fact that their hero, Mr. Carl, was gone. Some passed away as I mentioned earlier. One of our fine young engineers, Reggie Kerlin, who had been instrumental in the design of the airport, left to take a better position in Atlanta. We were pretty thin in engineering strength as 1970 rolled around but there was one important instance that bolstered me up and encouraged me to keep on keeping on. We had a fine engineer by the name of Pat Patillo in the firm. Mr. Pat had been a General in the Army, fought in WWII in Europe and was a civil engineer and a very close friend of my father. One day he came into my office and told me that he was one who was not going to leave. He told me that if the time came when only the two of us were left then he would help me turn out the lights. Mr. Pat deserves a lot of credit for G. W. Jones & Sons Consulting Engineers survival which is in its 123rd year. In fact, many employees added to my encouragement during that time. Tom Johnson, brother of "Pick" Johnson, was an outstanding employee and friend until his death on December 28, 1989. Lamar Speaks, Martin Phillips, D. J. Farley, Mary Ellen Mercer, Phoebe White, Ted Fountain, Luther Robinson, George Devaney, Jim Offutt, and Maureen Searcy were most encouraging in those days. My greatest encouragers and cheerleaders, however, were Libby and my mother during this time in our history. I came by each afternoon to visit with "Miss Betty" and she would advise and encourage me. Libby was the one who took it upon herself to be the stable, godly influence in my life that would see me through this most difficult period of time. Gradually because of "Miss Libby" and "Miss Betty" and a number of faithful employees, the office and farm businesses began to improve. I found that the engineering business, the insurance business, the abstracting business and the real estate business all rolled into one were not as difficult as the farming business. These office enterprises offered known input costs and receivables, educated personnel and an environment free from the weather. Farming offered none of these which makes the field of agriculture the most difficult of all enterprises because it is at the mercy of so many conditions beyond its control. Even though it was difficult, we continued into the decade of the 1970's pursuing the agricultural dream of Carl and Ed Jones that they started 30 years earlier.

Carl T. Jones
Soldier-Engineer-Business Leader

Chapter Twenty – Moving Ahead

The decade of the 1970's seemed to bring new life to both the office and the farm. Libby and I were blessed with a son, Raymond B. Jones, Jr., in November of 1969. Betsy had married Peter L. Lowe from Birmingham and Carolyn had married her high school sweetheart, John D. Blue, from Huntsville. We incorporated the office businesses under the name of G. W. Jones & Sons, Inc. in 1971 with Peter and me as principals and Carolyn as a minority owner. Peter had worked for the office for nine years and had attained an M.A.I. status so the appraisal and real estate entities were his responsibility. In addition to the farm, I continued to manage the consulting engineering business as well as the abstracting and insurance businesses. John Blue began work about this time at the office primarily working on real estate and appraisals.

Cattle and fescue seed were our main crops on the three large farms. We were still suffering, however, from the same malady with our cattle that had plagued us from the beginning of our fescue planting. In the months of May and October, our cattle did not do well. The cattle ran about a 3° temperature and generally felt bad. They didn't gain weight, breed or lactate well; they grew long hair and would stand in the water all day if it were available. The other months of the year they did well, which was baffling. To try and counteract this condition, we started a series of attempts to correct the problem. We planted 60 acres of alfalfa, 20 acres per farm, with the idea of hand feeding our replacement heifers. We were having very poor conception rates on our heifers and we

165

thought this would help. We soon learned that this did not help and after a few seasons we abandoned this idea. Eventually, we planted some coastal bermuda as well as wheat as we experimented on ways to stop this sickness in our cattle herd. Even as late as 1976, we were interjecting white clover in existing fescue pastures, trying desperately to find an answer to this cattle fever condition. The colleges, veterinarians and experiment stations were also at a total loss as the condition existed only on fescue pastures and was prevalent all over the southeastern United States.

Another effort to try to improve the efficiency of the cattle herd was to improve the overall quality by bringing in diversified genetics. We began to cross breed with Angus and Brangus bulls, especially on our first calf heifers. Calving had been a real problem with the Hereford breed; often 15% of the heifers needed assistance at calving. Today this is not a problem since we use EPDs (expected progeny difference) which can accurately predict calf birth weight as well as other traits. Breeders today tenaciously keep accurate records on the birth weight of their cattle so this trait as well as other traits can be used in bull and dam selection and thereby greatly reduce calving problems and improve performance. We continued to use these Angus cross bulls on our spring dropped cows until the mid 1970's. Calving problems got some better but it was not until we began using EPDs in the 1990's that a great difference could be noted. Another effort to improve genetics with our older cows was the purchase of superior Hereford bulls, mostly from Texas. Our search eventually led us to Montana and to a breeder by the name of Les Holden. We purchased a wonderful Hereford bull from him in 1970 by the name of C. Advance 601. When we brought this bull to Alabama, he was the talk of the cattle raising community. He weighed 2,350 pounds which was very large for this period of time and he was a perfect specimen. We eventually bred him to our top cows and saved his bull calves to further infuse his genetics into the herd. We even artificially inseminated about 100 of our cows and sold enough of his semen to pay for about half of the purchase price of 601. The genetics were good but his offspring seemed to be even more susceptible to the fever condition. We kept 601 until his death when he was about 14 years old. His progeny influenced our herd more than any other genetic influence until we began cross breeding with Red Angus in the 1980's.

C Advance 601
This Hereford bull was purchased in Valier, Montana in October of 1970. He provided superior genetics for the GW Jones & Sons Farm for many years.

The seed business was also undergoing drastic changes during this time. Labor to manhandle 25,000 bags of seed from the field was non-existent so we had to retrofit our field combines and our entire seed harvest effort to a bulk system. The field combine sack tying platform was replaced with a bulk bin with a bottom unloading door opening to the outside. A tractor would drive to this door with a front lift bucket and receive the load as the driver would rake out the seed into the bucket. The tractor would then lift and dump the bulk seed into a truck with 8' high solid side boards for delivery to the seed house. A suction pipe was installed in the winter of 1972 which was identical to a cotton gin unloading system. The seed would be sucked out of the truck and onto a horizontal conveyor belt and conveyed to the dryer bins. The next day after drying, the pipes were reversed and the seed sucked out of the dryer bins and deposited on the grainery floor to await the cleaning process. When harvest was complete, the pipes were reversed again and the raw seed were sucked into a mixing tank, ready for a trip through the cleaning plant. We went from counting raw seed from the field in bags to keeping up with it by the cubic foot. Most years we would account for about 50,000 cubic feet of field material.

Annually we sold approximately 800,000 pounds of clean Kentucky 31 Fescue seed. These were packaged in 50 pound bags, put on pallets and sold to seed brokers all over the Southeastern United States. Interestingly, the bags we used for our cleaned, ready to ship seed were purchased from the Werthan Bag Company in Atlanta, Georgia. If you have ever seen the movie "Driving Miss Daisy", then you would know that the story was about the Werthan family. "Miss Daisy" was played by Jessica Tandy, her son "Boolie" was played by Dan Akyroyd and her driver "Hoke" was played by Morgan Freeman. During those years we dealt directly with the Werthan family who always offered good service and had a good product. We still have a few of these bags on hand today and occasionally someone will see the movie and ask if they can have a bag with the Werthan stamp on it.

Another fortunate aspect of the 1970's was the addition of some very fine employees that came to both the farm and the office. Our current farm general manager, Roger Laster, was one of these that came on board in 1972. Roger came as a wonderful mechanic and maintenance man but has broadened himself into a valuable manager. Roger has a positive, cheerful approach to work that most men don't possess. The worst breakdown or problem becomes just another challenge to Roger who can fix most anything. Still another valuable farm employee that came in 1975 was Billy Bearden who served as the manager of the Guntersville Dam Unit. Billy was a "stand alone" manager who always got the job done even under difficult circumstances. Billy served as manager of this unit until his retirement in 2005. One incident that put a real strain on our cattle herd during the early 1970's, was a tremendous flood on the Guntersville Dam farm in 1973. Located adjacent to TVA's Guntersville Dam, the farm was inundated with water by a mistake of misreading a flood gauge system near Chattanooga, TN. This was before the days of the computer and the error caused about 20 feet of water above the normal pool level to be dumped on the farm. The result was that we lost 127 head of cattle and we had trucks and equipment that also flooded. Neighbors in their boats worked all day in the snow (early March) to help us ferry out and save 300 or more head. This incident caused us to back off and not use 500 or so acres of the most flood prone land. This land remains today in row crops along with several other fields that are also better suited for row crops than for pasture. At the writing of this book, the Guntersville Dam Unit stands alone as the best forage producer of the three main farms. The land above the flood prone area is very fertile and almost turns to grass as it continues to be our best beef producing unit.

In spite of the fact that we began to put some land at Guntersville Dam and Jackson County into row crops because of flooding, we were maintaining a herd of about 1,200 mother cows. Calves were born in the fall, and weaned the following June with the steers being shipped to Huntsville and kept until the following spring. The heifer calves were used as replacements for the herd and the balance of them sold to other breeders in the area. Additionally, our seed business continued virtually unchanged until the late 1970's when we began to also clean soybeans for sale. My work schedule remained in the same pattern. I would coordinate the farm activities early each morning and then spend the rest of the day at the office. Both office and farm personnel (almost 100 employees) were my responsibility. I dealt with hiring and firing, salary increases and personnel problems daily on a one-on-one basis. The hardest part of any business enterprise is dealing with personnel. Mine was one with a wide range of parameters. Some employees were without any education, some were highly educated registered engineers and many, especially in the early years, were older. Had it not been for several dedicated and wonderful employees, this part of my job would have been most difficult.

One morning I was very busy at my desk in the office when the receptionist buzzed me and said that James Leslie, who was still working at the farm, would like to see me. When James came in I could tell he was worried about something because his traditional smile was gone. "Morning James, what brings you to town?" I inquired. "Messer Ray I needs yo help." "What's wrong James?" "Well you see Messer Ray, I's on my way to court and I don't have nobody to go wif me and I's scared. I wants to know if'n you would go to court an just be there cause I bad needs a friend somewhere in that court room." "James I don't believe I would be any help just sitting in the audience, anyhow I'm awfully busy." Then James began to cry and I felt sorry for him so I agreed to go but only for moral support. On the walk of about two blocks to the City of Huntsville Municipal Court I learned that James had been drinking and was involved in an automobile accident. James had left the scene before the accident attending officer had arrived and the man he hit, even though his car was not damaged, was pushing the case against James. Sometime shortly after we arrived at the courtroom, the bailiff announced the case, "City of Huntsville vs. James Leslie." James, the police officer and the accident victim all went forward to Judge Bill Griffin's bench. The judge looked down at James and said, "Mr. Leslie, do you have legal counsel?" "How's dat, Judge?" James asked. "Do you have a lawyer to represent you?" responded the judge. "Aw yes sir, I's got a lawyer right over there, Messer Ray, he's my lawyer." The court room kinda snickered and Judge Griffin, whom I knew well, said "well Mr Ray, if you are going to defend your client you need to approach the bench." I knew then I had been trapped so I reluctantly proceeded to the Judge's bench. I couldn't think of much to say so I asked the judge for a five minute recess, after all I hadn't been on the case very long. During the five minute recess I asked James if the attending officer had seen him at the wreck. "Naw sir, he didn't see me because I done run off by then. That other man was there, but he be the only one that seen me", my client confided. When the court was called back into session the wreck victim gave his story and then the attending officer gave his report. I started our defense by asking the officer if he had seen my client at the scene. The officer testified that he had not seen James at the scene. At which time I turned to the Judge and said that it appeared that we have two testimonies that are completely in conflict. One witness says my client was there and my client says he was not there so we would have to ask for the mercy of the court since we lack uncontested evidence that my client was at the scene of the accident. Judge Griffin smiled and said, "James, I'm going to reduce the charge from D.U.I. to reckless driving and fine you $100.00 so you can keep your license. Assuming that's alright with you and your counsel." "Oh, yes sir judge dat's fine," James replied. "You do have $100.00 don't you James?" the judge responded. "Naw sir, I don't have $100.00 --- but Messer Ray does." James responded. Judge Griffin turned to me and said, "do you also pay fines, counselor?" "We always do judge," I answered as the courtroom exploded in laughter. On the walk back to the office I told James that I had attended my last court trip with him no matter how much he cried. James told me he wouldn't need me again because he was going to quit drinking. I didn't believe him, of course, but shortly after this incident James did quit and remained sober until he died about 20 years later.

Cattle grazing in Jones Valley.

Chapter Twenty One – Ventures and Involvement

The late 1970's and the decade of the 1980's found Libby and me busily raising our family and running G. W. Jones & Sons. The farm continued mostly on the same course of raising cattle and fescue seed. We involved our children in the farm activities as much as possible. 4-H Club Calf shows, working on the farm, herding cattle on horseback, gardening, etc. were the order of the day for all three of our children. Lisa and May, being older, were able to do some of the more physical labor. Lisa had a horse named "Denver" and May had one named "Cherokee" and they were called on regularly to help move herds of cattle. The girls were particularly good at moving cattle around so John K. and others at the barn were glad when they could help. Later Raymond helped with these moving duties and today manages the farm's cattle herd. Raymond's record keeping and breeding knowledge has transformed the current cattle herd into a class by itself in the commercial cow-calf world.

The club calf project of 4-H taught our children many lessons about life that we are most thankful for. Raising, breaking and training a 1,200 lb. animal so a 75 lb. girl or boy could show the calf is intimidating to say the least. We didn't win much when the girls showed but Raymond was fortunate enough to show the State of Alabama Grand Champion in his senior year of high school. We used these shows as a means to provide them money for college. At age six we opened each of them a "college fund" bank account. Into this account went the club calf money, birthday and grandmother money and the money they earned working in the summer. This established in their minds at an early age that they were going to college and that they would help pay for their education. When they did go off to college, Libby and I bought the clothes, cars, and gas but they paid for everything else. We firmly believe that they appreciated their education more because they felt they earned it and it wasn't just given to them. Education was not as expensive as it is today so each child had some money left when their college days were over. Once during Lisa's pursuit of a civil engineering degree at Auburn, she confessed to me that several times when she was tempted not to study, she would think about her sacrificial 4-H steers and then press on with her studies.

In the late 1970's, we were slowly going out of the fescue seed business. Lack of labor, a declining market and escalating costs were influencing the decision to reduce our fescue seed business. Soybeans and corn were replacing pasture acreage on the Jackson County and Guntersville Dam farms. Our seed cleaning plant was still intact so we decided to convert the grainery to clean certified soybean seed. This required a $40,000.00 investment consisting of an 80' elevator leg, 4 each 6,000 bushel soybean storage bins, spiral separators and retrofitting the cleaning process for soybeans. We raised soybean varieties by the names of Bragg, Essex, Lee, and Forrest which were all named for confederate generals of the civil war. Today these varieties no longer exist, having been replaced by large corporations who offer seed beans to farmers under a number. These numbers are usually correlated to a particular geographic area and whether or not the farmer wants early, medium or late beans. Our soybean venture caused a lot of extra work that first year in erecting the storage, cleaning and handling equipment. The 80' elevator leg and the storage bins caused the most involvement. We hired a crane to lift this elevator leg and its ladder and platform. The storage bins we decided to construct with farm labor. In retrospect, this was probably not a good idea because it was springtime and we were very busy. Grain bins are built with "rings" that are bolted together one ring at the time. The completed ring is then jacked up and then another ring is added until the desired height of the bin is reached.

To combat the labor shortage needed to bolt the bin rings together caused us to work at night. During the day the men at the barn would jack up the completed rings to the correct height necessary to bolt on another ring. Lisa, May, Raymond, Libby and I became the night crew as we bolted a new ring together after supper each night. I put a Coleman lantern in the middle of the partially completed bin which caused a stream of light to show through each bolt hole. May and Raymond, with their small nimble fingers, would then put in the bolts and start a nut on each bolt. Lisa, who was a little stronger, would tighten the bolts with an electric impact wrench. We could put a ring on in about two hours which left time enough to get to bed without being too late. It was a good solution to the problem and our 24,000 bushel soybean storage bin capacity was completed that way. The entire storage and cleaning system worked well and we cleaned, packaged and sold enough seed soybeans in the first year to pay for the $40,000.00 investment. We felt good that we were able to cover our initial capital outlay and involve our own land and existing equipment in the seed soybean business. At this juncture, we thought we had made an excellent decision.

In our second year, sales were not as good and we were battling the balloon vine. Balloon vine is a small "weed" that doesn't seem to harm anything except that it has two seeds in a small balloon shaped pod and that it grows to about the same height as a soybean plant. The balloon vine is a USDA noxious weed, the seed of which has exactly the same properties as a soybean seed. It is round, rolls in the spiral separator at the same rate of speed and was impossible to separate from the soybeans with our cleaning apparatus. Our only defense was to "hand pluck" balloon vine from the field. This was very expensive and was one of the reasons that our net revenue dropped to $10,000.00 in the second year. We abandoned our seed soybean venture at the close of our fifth year.   We didn't lose money but we were not able to overcome the balloon vine and the intrusion of the large corporations into the soybean seed business.

The large corporations bred superior genetics into their soybean varieties and they bought up the rights to the "confederate general" named varieties. These new varieties became superior and farmers today prefer them to the original ones. By 1988, we had disposed of our last soybean and had discontinued saving fescue seed. The once proud seed grainery and its cleaning equipment stood silent, providing only a reminder of another way of life and employment for hundreds of people. The seed grainery had served G. W. Jones & Sons Farm for almost 40 years. Today the cleaning apparatus is still unused, some of the soybean equipment has been sold and the sturdy oak structure built by

Carl and Ed in the early 1950's is mostly used for the storage of equipment connected with the current farming operation.

In the midst of all these changes on the farms, our "office" businesses were doing well and were profitable. I was involved in several community as well as farm organizations. I served as president of the Madison County Cattleman's Association in 1957 and the State of Alabama Cattleman's Association in 1976 as well as being inducted into the Alabama Livestock Hall of Fame in 1988. On the community front, I served as president of the Huntsville Rotary Club, Chairman of the Alabama Conservation Advisory Board, Auburn Alumni Association Board, Chairman of the University of Alabama Huntsville Foundation Board and a host of community and civic involvement committees. These all helped our businesses and even though they were time consuming, I was able to manage them in a way so that I had time for Libby, the family and church activities.

I served as a Deacon of the Mayfair Church of Christ from the early 1970's until 2003. The church has been at the forefront of our family's life throughout our marriage. Libby and I have taught bible school, served on committees, gone on mission trips and numerous other efforts at Mayfair during our married life. We consider the church to be the most positive influence on our family. One very meaningful event related to the church came in 1983. Willard Collins, the then President of David Lipscomb College, came to Huntsville and asked me to serve on Lipscomb's Board of Trustees. I was very busy at the time and prepared myself to say "no" to the invitation. Willard Collins though had a very persuasive personality and I was infatuated by him and agreed to serve. My tenure at Lipscomb lasted over 20 years, during which time I served with four different presidents. Today, Lipscomb is a University and has increased its student enrollment by almost 1/3 to over 3,000 students since my first meeting in 1983. Fortunately, all of our children and their spouses have attended this small religiously oriented school in Nashville, Tennessee. Currently, our two college age grandchildren, Elizabeth and Allison Yokley, are attending Lipscomb and they are enrolled in a nursing curriculum. Needless to say that visit from Willard Collins was most meaningful to me personally and also to our family. It was through Lipscomb that I was fortunate enough to forge lifelong friendships with some of the finest men I ever met. The Lipscomb experience also exposed me to a level of financial involvement that I would never have known even existed as I served the school as its Real Estate and Finance Chairman. Each time I returned from those meetings I felt inspired, both spiritually and secularly.   The Lipscomb experience occupies a special place in our lives and has truly been a blessing.

My mother's insistence that I go to a wedding and meet that school teacher and Willard Collins' insistence that I join Lipscomb's board stand out very vivid in my life. We all pray for guidance from God and many times we entertain "angels unaware" and I believe this happened in these instances and others in my life. There is no way that a country boy, like me, living on a gravel road in 1939 and starting out with no electricity should have had such opportunities without Divine guidance.   I've rubbed shoulders with men of enormous character and talent in my life and I'm most thankful for each one. Some of those men came through my association with the cattle business. Through agricultural circles I met men like Dr. Tom Haggai, Jerry Clower, and Dr. Charles Jarvis who were all accomplished nationally known public speakers. Several governors, U.S. Senators, three American Presidents, several University Presidents and numerous hard working farmers and cattlemen that I consider to be equally as great have crossed my path. Only in America can stories like mine evolve; my hope and prayer was the same in the mid 1980's as it is today, simply that God will continue to bless our great nation.

May Criner Jones and her 4H club steer. This photo made the cover of the Alabama Cattlemen's magazine.

Chapter Twenty Two – The Mystery is Solved

The early 1980's found the farm still fighting the cattle fever condition and Libby and I seeing our family growing up. Lisa was in college studying civil engineering. May was in college studying marketing. Raymond was in high school and all of our children worked in the business during the summer of their grade school and college years. Lisa worked at the office helping with engineering projects and she helped with the cattle when needed. May made a big garden in which she produced, picked and delivered fresh vegetables for our office customers each week. Raymond was involved in the farming operation and was learning some valuable lessons about the field of agriculture as well as hard work. Also involved as summer farm workers were several cousins and nephews of the family. Philip Bradley probably worked the most summers as well as total time for the farm. Philip today is the pulpit minister for the Church of Christ in Guntersville, Alabama. Philip was the one who always kept the crew laughing; the work was never too cold, hot or dusty for him to be the hardest and most gregarious worker. Others that also worked on the farm during this time were John Wallace Blue, Steven Mercer, Joe Strickland, and David Strickland and each would come and go as their school and or college schedule would allow.

Ruth Jefferson, our wonderful housekeeper, came to our family the year Raymond was born (1969) and she and Libby would cook lunch each day for the working kids. The highlight of the day was lunch. Sometimes I would drop by just to hear all of them talking and laughing at the same time. Lisa and

May, along with the boys, made for a lively bunch around Ruth's and Libby's good food. This was a time when they would "mimic" their parents or other workers to the laughter of all. One day I slipped in the back door at the same time David Strickland was mimicking me so I waited quietly until his performance was over. When I made myself known, the whole room, including Libby and Ruth, came unglued and they told it over and over for days.

One of the funniest stories about this time involved Philip, who was working one winter between quarters while in college. It was a real cold day with the temperature in the 20's and John K., Phillip and crew were feeding the cattle in the pastures. Philip got out of the truck to unlock the gate and found the lock frozen. Not to be denied, Philip began to blow his breath and squeeze the lock to thaw it out. The harder he tried the closer he came in contact with the frozen lock. Finally the lock was opened, thanks to Philip's intimacy with the frozen obstacle. Meanwhile, John K. had told the other helpers in the truck to be quiet and let him play a trick on Philip. Shivering, Philip got back into the truck. John K. said, "Philip that lock was pretty frozen wasn't it?" "It sure was, I almost had to get it in my mouth before I could unthaw it, he replied." John K. said, "You know I had trouble with that same lock yesterday and I had to pee all over it before I could open it." At this point Philip began to cough, spit and breathe hard until the rest of the crew couldn't hold their laughter. They laughed so hard that Philip knew he had been had. The entire farm community enjoyed this story for the rest of the year.

Looking back, I believe this segment of time was a golden period for us. Libby and I were healthy and working with the

Lord trying to grow these kids as well as our crops. I'm afraid that many times in life we pass along too fast to experience the beauty and special times that occur along the way.

The kids were always raising animals of all sorts. We had chickens, quail, ducks, geese, a goat, a fox, calves, and deer as well as the usual dogs and cats. One year Raymond raised about 30 Broad Breasted Bronze turkeys and gave them to his aunts for Thanksgiving. Some of these turkey weighed in the 40 to 50 pound range and were delicious on the Thanksgiving table. The kids all enjoyed one turkey gobbler in particular who would attack anyone that passed by. Philip got attacked one day, much to the delight of everyone, and a battle royale ensued. At the end of the fight, the turkey had a crook in his neck that lasted until he was prepared for the Thanksgiving table. On another occasion we raised a goat with the dog in a dog pen. The goat grew up thinking she was a dog. When we went walking she would run and jump and do everything the dog would do. She was a good pet until one day she ran into the house, jumped up on the couch and began eating Libby's draperies. After this we gave the goat away. On still another occasion, Raymond raised about 20 Canadian geese that would follow Raymond anywhere he went. The geese "imprinted" themselves to Raymond and would follow him by the sound of his voice all over the place. We thought this was unique so one day as Libby was having a big tea party we decided to show off the geese. The tea party was in full swing when we entered the parking lot and patio with the geese following Raymond. The women thought the geese were also unique and real pretty. Everything was working fine until the geese decided almost simultaneously to use the bathroom. Goose droppings were all over the patio and parking lot and

"Miss Libby's" wrath caused all of us to make a hasty retreat. We all laughed about it later and in spite of the conditions, I believed the tea party guests also enjoyed the event. Lots of stories still survive about the menagerie of animals that the kids raised on the farm and are retold to the grandchildren regularly.

Also during the early 1980's, several southern universities were working on the fescue fever problem primarily based on a farmer in Mansfield, Georgia noticing a strange phenomena on his cattle farm. He had several KY 31 Fescue pastures and noticed that on one particular pasture his cattle looked healthy and were gaining weight. On another pasture his cattle had long hair, looked unhealthy and were losing weight. After calling in his veterinarian, they discovered that the healthy cattle had normal body temperatures and the unhealthy cattle had an elevated temperature of 2 – 3 degrees. Scientists at the U.S. Department of Agriculture in Athens, Georgia began investigating this puzzling situation. Pasture grass at this time consisted mostly of KY 31 Fescue and occupied 35 million acres in the nation and economically was of great interest to the cattle industry. Several land grant universities began grazing trials trying to understand what was occurring with this popular pasture grass which caused the fever in cattle and which was becoming known as "Fescue Toxicity".

Auburn University was also engaged in the effort to understand what was happening and causing "Fescue Toxicity". One day I got a call from Aubrey Smith inviting me to come to the Black Belt Experimental Substation near Marion Junction, AL where he was the station manager. I knew Aubrey Smith through the Alabama Cattleman's

Association so I visited him with the idea of seeing first hand a discovery he had made involving fescue toxicity. I observed 10 different 10 acre fields planted in KY 31 Fescue that had steers grazing on them. Fields 1 – 9 had the cattle with the usual fescue toxicity symptoms on them while field 10 had cattle that showed no symptoms of the toxicity. In relaying his story, Aubrey told me that fields 1 – 9 were planted in September with seed harvested earlier that year. Field 10 was planted with some old seed because they ran out of seed planting fields 1 – 9. Unknowingly, Aubrey had hit at the very heart of the problem. Eventually and after many experiments, a fungus was discovered within the fescue plant. This fungus was subsequently identified as "Acremonium Coenphialum" which resides between the cell walls of the plant. The fungus is seed transmitted only and lives in plant stems, leaves and seed, probably stimulating the plant to produce a substance which is highly toxic to livestock.

Aubrey's discovery caused quite a stir within Auburn University Agricultural circles. One of the first efforts to quantify the problem was to establish a fescue fungus diagnostic laboratory. Auburn asked me to head up an effort to fund this laboratory so farmers could at least know the level of toxicity in their pastures. In raising funds for the laboratory, we asked each cattleman to donate $1 for each owned acre of fescue pasture. We had a good response to the request and raised about $30,000.00 to kick off Alabama's diagnostic laboratory at Auburn. A lot was learned from this effort by farmers seeking answers to the fescue toxicity problem. Cattle on fungus free pastures did much better than those on fungus infested ones. Aubrey's Black Belt Substation did subsequent grazing trials and found that steer performance jumped from .65 lbs of average daily gain on fungus infested pastures to 1.48 ADG on fungus free pastures. The goal now for the cattleman was how to manage around this fungus problem on their permanently established pastures. The cattle industry finally knew the cause of this "fever condition" and its effects and everyone connected with these 35 million acres of KY 31 Fescue sought answers for the future.

Armed with knowledge of this discovery, the farm launched several efforts to try to diminish the fungus and its toxicity level in our pastures. Our first try was to reestablish our pastures with "old" seed. It seems that the fungus dies out after about a year in stored seed. Since the fungus is not contagious and is entirely reproduced through the seed, it seemed logical to just plant "fungus free" seed. The diagnostic laboratory stayed busy running tests on pasture grass samples from farmers measuring their percent of fungus infestation present. This worked well until the newly established fungus free plant reached maturity and drought or other stress intervened. There was something about the fungus that strengthened the fescue plant. These fungus free plants rarely survived their third year of life. Still another idea was to interject legumes (clover) into the existing fungus infested fescue pasture. This had some reward but was only marginally successful. The seed industry moved quickly to develop "novel" varieties of fescue that would hopefully replace the nation's acres of KY 31 Fescue. There were several new varieties that hit the market in the late 1980's and early 1990's. Most of these "novel" fescue varieties were ineffective or not economically feasible. The Pennington Seed Company finally developed the most successful non-toxic variety to date. The trade name of this variety is MAX-Q. MAX-Q has been used over a wide area in the south and offers non-toxic forage and is fairly resistant to drought and disease. Currently the farm has about 400 acres of MAX-Q and it is doing a good job for us. MAX-Q is not quite as productive in total volume as the old fungus infested KY 31 Fescue, but is of higher quality. Cattle prefer MAX-Q forage and of course their performance is much better. The main drawback is the cost of the seed at planting. MAX-Q seed is about five times more expensive than the old fungus infested KY 31 Fescue seed.

The farm has been more productive and easier to manage by just knowing the cause of the fever condition in the cattle. Merely rotating the cattle from pasture to pasture during the months of heavy infestation has helped. Supplemental feeding and less fertilizer at certain times of the year have been of some benefit. Our cattle are becoming heavier each year and our calving percentage greater by knowing the cause of the fever condition. The entire cattle industry owes a sincere debt of gratitude to people like Aubrey Smith, Dr. Don Ball, Dr. Garry Lacefield and many other agronomists who helped solve this problem. Dr. Don Ball of Auburn University has been a good friend and regular source of advice for the farm's pastures as we have struggled to overcome this fungus problem. We currently believe we have a handle on the fungus condition. We only wish that Carl and Ed Jones were here to observe cattle that are as healthy as they are today.

The fever condition that had plagued us from the very first planting of KY 31 Fescue in the late 1940's had now been finally identified. Fescue Toxicity and its conditions are now known to the cattle raising industry where cattle graze on millions of acres of fescue in the southern part of our country.

Newer varieties of fescue and diverse management procedures in the future will help restore this "wonder grass" to its rightful place of significant importance as a forage crop. Ed and Carl recognized its value when they first saw it growing in Menifee County, Kentucky in the 1940's and they had the foresight to base the farm's production on this one plant. Even though that production has been curtailed by this "fever condition" over the years, all of us could safely say that we have been pleased with KY 31 Fescue. Additionally, all of us from Carl and Ed until today could also safely say that we are glad that this mystery of fescue toxicity has finally been solved.

Newly established fungus free pasture in Jones Valley.

THE FARM
IN JONES VALLEY

PART III

Chapter Twenty Three – Changes

The decade of the 1990's brought significant changes to the farm as well as the entire G. W. Jones & Sons family. Lisa had married Mark Yokley in 1984. Mark was a farm raised boy from Edmonton, KY and he and Lisa met while attending Lipscomb University. They transferred to Auburn University in 1981 with each of them pursuing a civil engineering degree. Lisa decided as a sophomore in high school to become an engineer. Upon graduation in 1985, they moved to Huntsville and began work at G. W. Jones & Sons. The two of them breathed excitement and life into the 99 year old firm that my grandfather started in 1886.

In 1986, the engineering firm along with the farm and all of the other entities, celebrated its 100th anniversary. Huntsville's Von Braun Civic Center was rented and we culminated a week long celebration with a 500^{+} person dinner at the VBCC. All the living descendants of G. W. and Elevena Jones were seated at two head tables along with several long termed employees. Dr. Willard Collins, president of Lipscomb University, led the invocation and Louis Salmon, our attorney and good friend, was the featured speaker. After dinner the guests were moved into a large adjacent room for the purpose of viewing a multi-media presentation relating the history of the G. W. Jones family from 1804 to 1986. Pamphlets of this history, which was the basis for the multi-media presentation, have been widely used by the firm over the years as we called on potential customers. Employees of the firm dressed in 1880's clothing all week and had an open house at the office. We received a lot of compliments and good will for the firm from friends and clients. Lisa and Mark, fresh from college, added a lot to the planning and logistics of the event.

Mark had worked with the land surveying crews during his summer breaks while at college which gave him hands on experience in the field which would prove very valuable for the future. Mark quickly related to everyone in the organization from the field crews to the engineering design personnel. Being a farm boy, he was very much at home in the field and quickly became a favorite with the crewman. Once, while on a survey, the crew came across a persimmon tree full of green persimmons. Of course there is nothing more sour to the taste than a green persimmon. Mark and one of the crew men picked a few and put them in the cooler with the rest of their lunch. When lunchtime came, Mark got out the chilled green persimmons and put them around his plate of food. One of the crewmen asked him what the pretty green fruit was that he was having for lunch. Mark, without batting an eye, told him that they were "French Plums" and offered him one. The fellow took a bite and put on quite a show of spitting and coughing. The other crew members enjoyed the event and Mark became the subject of the ruse all over the firm.

May married Mike Patterson in 1987. Mike was raised in Knoxville, TN and began his college career at the University of Kentucky as a scholarship football player. Mike eventually left Kentucky and enrolled at Lipscomb University where he met May. May received a degree in marketing and Mike a degree in accounting and upon graduation in 1987, they moved to Huntsville. Mike began his career with a local C.P.A. firm and eventually earned his personal designation as a C.P.A. May went to work for G. W. Jones and Sons and became a licensed realtor. May, likewise, was a big help with the 1986 centennial celebration and the youthful energy of these three additions to the office was most helpful.

Meanwhile, back at the farm, several changes were occurring as we neared the decade of the 1990's. In 1986, the City of Huntsville petitioned the family to build a five lane road across the farm connecting Whitesburg Drive and Bailey Cove Road. The farm donated 27 acres to the city for this road which was eventually named Carl T. Jones Drive. The road had quite an impact on the farm because it severed several hundred acres south of Carl T. Jones Drive from the main part of the farm. The following year after the road was constructed, the farm developed Jones Valley Gardens. May, being a realtor, was put in charge of marketing these lots which became very popular. One unique feature of the subdivision was that as developers we established all the landscape plantings before the lots were sold. Decorative lighting, green areas, rock walls and wooden fences were also completed before offering the lots for sale. Subsequently, we also developed Jones Valley Gardens Phase II on still another portion of the land severed by Carl T. Jones Drive. Other developments precipitated by the road on the farm were a retirement community named Somerby at Jones Farm, Southwood Presbyterian Church and the Mayfair Church of Christ. Even though these developments occurred over 15 years ago, one would not have to look very hard at the architecture and landscaping of these developments to realize that the intent of the family was to maintain the beauty of the valley. The farm in Jones Valley has been our home since 1939 and hopefully will be for many years to come and we sincerely want to preserve its beauty. Even though we came to this beloved valley to farm, we realize that changes will inevitably occur. At the writing of this book, 2009, we are still fully involved in the cattle business. All of our neighbors have sold out and moved on but the farm remains a viable enterprise and today is almost surrounded by the growing City of Huntsville. We are told that the farm that Carl and Ed struggled to establish in 1939 is today the largest urban farm in the nation.

The rain came in torrents on January 3, 1991 as we buried "Miss Betty" at Maple Hill Cemetery. The day before during the visitation at her home on the farm it also rained. The temperature was warm for that time of the year and literally hundreds of people attended the visitation. A line of visitors under umbrellas stood in the rain on the road from her house on the farm to Garth Road as they paid their respects to this great lady's family. This wonderful person who was beloved by everyone who knew her, was now gone. Her pioneer spirit held the farm together during the war years. It was "Miss Betty", more than anyone else, who deserved the credit for the farm's survival. Someone at the wet funeral said that they believed that even the angels in heaven were also weeping for this great lady. I still have the hundreds of cards that were sent to the family in the days following the funeral. Almost 90% of the cards used the word "lady" when referring to "Miss Betty". A small lake, pump house and garden was later dedicated and named for "Miss Betty" at the intersection of Carl T. Jones Drive and Garth Road. A small but fitting recognition of this lady's dedication and perseverance to a farm located in a valley that bears the family name.

The office was expanding about this time and more space was needed to accommodate the growth. A larger building one block south of the location of the original office was remodeled to meet the needs of the growing engineering business. Libby came up with a suggestion that we eventually used for a final floor plan. The flow of personnel and foot traffic based on her suggestion has worked well over the years in this building as it has served the engineering business. I moved into this new building at 401 Franklin Street with the engineering business. Peter remained at the original 307 Franklin Street address with the real estate and appraisal businesses. Both locations kept the name G. W. Jones and Sons that had existed on Franklin Street for over 100 years. The move proved to be a good one, offering more space and opportunities for the growing engineering firm.

The early 1990's saw the farm gradually phasing out calving our heifers in the spring. We found that with a little more feed and size selection we could have our heifers calving in October as two year olds rather than in the spring as 2 ½ year olds. This not only provided a longer productive life for the cow, but solved a lot of management and marketing problems. Our cow herd consisted of over 1,000 females at this time, two hundred of which were first calf heifers. The Santa Gertrudis bulls were phased out in 1990. Red Angus bulls were being purchased as our breed for the future. We liked the Red Angus calves from the beginning because they brought a lot of desirable traits to their offspring. One of the real advantages was on our first calf heifers; the Red Angus Breed is inherently smaller at birth. Still another advantage was the fact that the Red Angus breed is "polled", meaning that they have no horns. We have continued breeding a lot of Red Angus up to the present time and each year the herd has fewer horns and less white on their face and body resulting in less eye problems.

Traditionally, up until the 1990's, we had sold our steers at about 18 months of age weighing around 900 pounds. This practice was abandoned in the early 1990's to provide more acreage for fall calving mother cows. It was expensive to carry these big steers over a second winter and our best profit came from a 7–8 month old steer sold directly off the cow. One instance I remember about those big steers came on November 15, 1989. On this date, Huntsville experienced a deadly tornado which killed 35 people and caused severe damage to the southern part of the city. With no warning the tornado first touched down on the west end of Airport Road and traveled east right down the middle of the road to Whitesburg Drive. The tornado lifted up at this location but set down again on Jones Valley School and traveled east across the northern end of the farm and through Greenwyche Village before leaving the valley. Fortunately, school was over when the tornado hit and only about 15 children were present at the time of the late afternoon touchdown. Miraculously, three teachers were walking the children down a hallway next to a stairwell to get refreshments when the school roof collapsed. There were also three painters painting the stairwell at that same exact instant. Quickly the teachers and the painters herded the children into the stairwell for their safety. By now it was dark and when the building quit falling around them, they exited the stairwell and the demolished school and went across Garth Road and into the farm pasture. This was the only open area and must have felt safer for them at the time. When rescue personnel found the children in the pasture, the three teachers and the three painters were still huddled over them and none of them were hurt. The school was completely destroyed and afterwards no one could imagine how anyone could have lived through such destruction.

The farm lost several cattle, mostly trauma related injuries from flying debris. A group of these big steers were in a pasture right across from the school that we called the "pecan orchard" pasture. The pasture had about 20 very old and large pecan trees that were in the path of the tornado. These trees were all destroyed and the flying limbs from them killed several of the steers. Tin, shingles, paper, garbage cans, plywood and most anything else one could name was scattered in our pastures. The cattle that survived looked like they had been rolled around in the mud and they were real sore and stiff for several days afterward. As we began treating abrasions on the steers, we found a lot of pencils stuck in their bodies from the school. One school paper item was found over six miles away from the school. The strength of this storm was devastating to our community and the farm expended a lot of effort and money dealing with its effects. Thankfully none of our family members or employees were hurt. The farm was cut off from town as far as electricity, travel and communication was concerned for several days but nevertheless we were thankful that our lives had been spared. In the weeks following this devastating tornado, there were hundreds if not thousands of acts of kindness among our citizens toward each other. The storm had brought our community to its knees and from our knees many prayers were lifted up to the Heavenly Father in the days following this devastating event.

This period of time involved quite a number of changes, some good and some bad. As I mentioned earlier in chapter 19, the one "constant" in life is change. Most of the changes that had occurred, were occurring and would occur to the farm were normal and should be expected. Farming is a metamorphis of changing events that occurred in the past as well as those occurring in the current year. No two years are the same for the farmer. His job is to accept these changes, plan ahead and pray for guidance as he approaches each season. Challenges for the farm have been and would be no different.

Carl T. Jones Drive

Chapter Twenty Four – New Leadership

The farm received a real infusion of youthful leadership in 1993. Raymond had completed his college work and returned home to the farm, a move which would significantly add to the farm's efficiency. Raymond had received a degree in marketing from Lipscomb University in 1992. Following graduation from Lipscomb, he enrolled in Auburn University in the school of agriculture as a graduate student "special ten". This "special ten" designation allowed him to take a variety of agricultural courses, the prerequisite requirements of which were waived by the dean. This was not a degree granting program but it allowed him to take a wide range of subjects. Subjects such as forestry, beef cattle production, real estate and finance, forage production and others were taken as a "special ten" graduate student at Auburn. These courses prepared and focused Raymond's education on future needs of the farm and G. W. Jones and Sons. The U.S. Armed Services were downsizing in the early 90's so Raymond was not required to serve in the military and as a result he came straight to the farm from college. I immediately began to shift the leadership of the day to day farm activities onto his shoulders. Raymond had worked on the farm off and on since he was about age seven so he was familiar with much more than he realized when he started. Within a few years of Raymond's return, we had phased out spring calving. Spring calves didn't fare as well as fall calves and they were not as profitable, primarily because of the summer heat. We had begun to use "balancer" bulls which were a cross of the Gelbvieh and Red Angus breeds. A one-half Gelbvieh and a one-half Red Angus bull bred to our Hereford x Red Angus cows produced a one-fourth Gelbvieh, one-fourth Hereford and one-half Red Angus calf. These percentages remain close today but are gradually moving toward being a higher percentage of Gelbvieh. This cross has proved to be a very efficient one for us as well as customers buying our cattle. Recently we had an occasion to feed out a load of heifers with a feeder in Iowa. We had these calves graded while hanging on the rail after slaughter. Surprisingly, they graded 95% choice – prime and had a 67% cut out value. These statistics confirmed the fact that we were on the right track in producing commercial animals for the feedlot. Additionally, we have had numerous comments from our customers complimenting this breed cross. One customer has bought our entire steer calf crop for the last six years so he must be pleased with them. I'm sure we will vary our breed cross again in the future, but we have been very pleased with the Red Angus and Gelbvieh cross that Raymond helped to initiate in the 1990's.

The mid 1990's saw the farm phasing out of the cattle business on the Jackson County Unit. Mike Miller, son of Leon Miller, had worked on that farm from the time he was a small boy and had been selected as Jackson County's manager in 1985. This farm was fast becoming a row crop / timber and wildlife unit. Mike made the transition well and has subsequently planted over 125,000 trees in wildlife corridors and / or CRP land. Some old growth timber cutting was recommended by our conservation wildlife advisors which provided openings in the heavy mountain timber. Additionally, Mike has constructed approximately 35 miles of mountain roads, several hunting houses and sheds as well as a duck pond. The trees consisted of pine, poplar and oak varieties of cherrybark, shumard, nutall and white. Green patches were established throughout the farm as well as autumn olive and chickasaw plum thickets. Mike Miller has close to 30 years service by the time this book was written and has been a wonderful manager for the farm. Mike and his family live in a house adjoining the farm and have managed the place well over the years. He is always pleasant, honest and exhibits a Christian attitude toward his job and the people he meets. Mike is another in a long line of faithful employees that deserve much of the credit for G. W. Jones & Sons Farm's success.

Roger Laster was elevated to the position of general manager in 1996. Roger had worked for the farm since 1972 and had become a very valuable leader. Roger has a mechanical talent that can fix almost anything. In addition to this talent, he is a good leader of men. He is always jovial and pleasant and the men working in his crew are noticeably at ease and gainfully employed. Roger has seen the transition from the fescue seed harvesting days to one of a total cattle enterprise and then to a partial row crop operation. Roger saw and implemented the seed cleaner and grainery equipment as it was transformed from fescue seed to soybeans. Haying, mowing, fencing, fertilizer procedures and equipment usage were mostly devised by Roger. Recently, Roger has had to learn more about the cattle business. He was trained as a mechanical man early in his career so the cattle business was a new experience for him. Roger has successfully grasped every phase of the cattle business from calving to marketing. Roger likewise is always a pleasant person and fair with his employees while demanding from each man a full and hard day's work.

The leadership of these men during the 1990's and into the current century has been very meaningful to the farm. The leadership team comprised of Roger Laster as general manager, Billy Bearden as Guntersville's manager and Mike Miller as Jackson County's manager was very successful. Farm work on all the different farms was and is a team effort. It is not unusual for personnel from each farm to travel to the farm effort that needs help. This is particularly true during cattle working time. The working pens, trucks and trailers are manned by personnel from all the farm units for the task at hand. Calving time, planting time and haying are all considered to be everyone's responsibility, no matter the location. Another philosophy from "Mr. Carl", "it is much more effective to have a team approach than an individual approach to things".

Another effective addition that Raymond brought to the cattle end of the business was getting all of our cow records on the computer. Since the early days of our being in the cattle business, we have had a numbering system on each individual animal. Carl and Ed initiated this in the early 1950's. Their numbering system was rudimentary and really had no final focus of use. They were so intent on a numbering system that in the very early stages they "hot branded" numbers onto the cows. This was not very meaningful because the branded numerals would "scar" over time and become illegible. This practice was eventually abandoned; however, holding brands remained until the turn of the century. Alabama, like most cattle producing states, has a registered brand law which allows each cattle breeder one registered brand to be listed in his name. The farm's original registered brand was a JB which stood for the Jones Brothers of Carl and Ed. This was later replaced by a single J. The B would eventually blotch out with scar tissue and the branding iron itself was heavier and more expensive. The J brand worked much better and is still registered to G. W. Jones & Sons Farm today. Branding on the farm has been abandoned in recent years and replaced by the less intrusive method of ear tattooing and ear tags.

Ray and Raymond Jones examining computer records of cattle.

Steer calves are tagged at birth with yellow tags and heifer calves with white tags. A corresponding number is permanently tattooed in the left ear of each calf. Each calf is weighed at birth and the bull calves are castrated with an elastrator band. At birth the mother cow is always present as she looks after the newborn calf so her number as well as the date and the birth weight of the calf is also recorded. This data is put into the computer by Raymond and attached to the cow's permanent record. The computer has vastly streamlined the farm's cattle operation. No longer do we wonder how many cattle or which cattle are in a particular herd. The computer won't let duplicate numbers be entered and the flow of information about the entire herd is accurate and readily available. Weaning weights and other information can also be attached to the mother cow's permanent record. The farm weans its calves in June of each year. The previous fall the calf is dropped within a 63 day period (Oct. 1 – Dec. 2) which, sizewise, makes for uniformity at weaning. Weights at weaning are from a "load bar" scale which also records data via a computer. Currently, Raymond is using "Red Wing" software and the innovation of the computer world has helped our cattle business as much as anything else. Field data collection in the old days was taken via horseback from a roped and tied calf. We abandoned this method several years ago and catch each calf by hand using a pickup truck. This works pretty well, particularly if the calf is only one day old. The main obstacle is the mother cow who usually doesn't like for her calf to be caught. Sometimes she becomes defensive and will put the men back in the truck. Several times during the calving season the men will have to retreat up into the pickup truck bed to tag the calf. I have known some mother cows to jump up into the truck bed after a bawling calf. Most mothers all over the world are like that when it comes to protecting their babies; what a blessing.

In the mid nineties, we began looking for a more sophisticated way to market our steer calves. Raymond had made several trips to Montana and Nebraska to purchase "balancer bulls" which were half Gelbvieh and half Red Angus. This cross, as I mentioned earlier, was a good fit for our herd and we were producing outstanding calves. At the time, our marketing plan was to sell our calves right off the cow in June. This, however, didn't take advantage of our genetics. Subsequently, we placed our calves in feedlots in Texas, Colorado and Iowa and over several years we learned a lot about feeding our cattle to a finished product. These trips west were mostly successful, however, by the turn of the century this practice was abandoned. High fuel costs and a significant customer interest for our calves right off the cow in June helped make the decision to revert back to our previous method. Heifers from these matings have always been in demand from local cattlemen as they would expand, upgrade or rebuild their herds. The farm usually kept about 1/3 of the heifer calf crop for its replacements. The computer helps us make selections on which of these heifers to retain. We retain the best of our heifers and mate them with the best "balancer" bulls we can purchase which produce outstanding calves each year. We market the steer calves via a telephone auction. On a given day in the spring, we will have the current interested buyers call in at a set time and we then proceed to auction the steers in truck load lots (50,000 lbs). This method has been very satisfactory to both the farm and the steer buyer. The buyer is spared travel costs and the farm consistently receives a premium over the national market.

I think Carl and Ed would be very pleased with the cattle product we are now producing. I know they would embrace the current numbering system, computer use, breeding and marketing methods that have evolved as we enter the 21st century. I wish I could call them and the hundreds of cowboys that have worked on the farm with the cattle herd since 1939 back for a visit. I would like to conduct a tour for them; much like some of our visitor tours, showing all that has been accomplished. In all candor though, I probably would not be able to conduct the tour for being overcome with emotion.

These very men that I have known and that had worked so hard on the farm occupy a very special place in my heart and life. My life has been spent with them and these farms. I could have done other things with my life rather than till the soil and raise cattle, but as I look back through the sweat, cold, dust and tears, I wouldn't change a thing. I would only want to be able to spend more time with those wonderful dedicated people who have worked so diligently under the banner of G. W. Jones & Sons Farm.

Chapter Twenty Five – Wildlife

Historically, agricultural land values have been derived by determining soil types, crop productivity, elevation above flood, slope and standing timber. Farm owners have become very familiar with these characteristics, particularly when buying or selling a tract of land. About 25 years ago, another value crept into the evaluation of agricultural tracts of land, especially tracts that were timbered. This relatively new delineation of land value is the ability of the tract to support wildlife. Wildlife has surfaced today as one of the most important aspects of determining rural land value. One of the first questions asked by a prospective buyer is usually, "what wildlife species are prevalent on the tract of land?" Large game animals in Alabama, deer and turkey, are the most widely sought after. Migratory birds such as ducks, geese and the mourning dove are also considered. Small game such as quail, rabbit, squirrel, and raccoon are less often inquired about.

In our area of North Alabama, deer and turkey were non-existent in the mid 1960's. The Alabama Department of Conservation made a concerted effort to re-establish deer and turkey in all 67 counties and especially those not supporting huntable numbers. G. W. Jones and Sons Jackson and Marshall County farms were chosen as live-trapped release sites by the Alabama Department of Conservation for this effort. We had to agree not to hunt anything for five years while the deer and turkey were re-establishing themselves. This was an exciting time for me as I was asked to help with the trapping and release of these new wildlife residents to our area. The deer were trapped in baited box traps in south Alabama on several different tracts of land where the deer were over populated. The deer were then hauled in a padded box to the release site. In the winter of 1964, we released about 30 head of deer on both the Jackson and Marshall County farms. The deer from the first season on seemed to like their new home and began to multiply. It's been over forty years since this release and the deer continue to increase and spread to adjoining areas. In some areas, the population is excessive and the deer cause a lot of crop, garden and yard damage. Huntable numbers after that five year period were sufficient in all 67 counties and today Alabama has a very liberal hunter limit in an effort to control the sheer numbers of the expanding deer herd. Farms that have a good deer population are considered more valuable than those with inferior numbers. Deer hunting is very marketable to the large population of urban hunters that want an outdoor hunting experience. The annual deer kill in Alabama is estimated at about 450,000 which produces approximately 22 ½ million pounds of venison each year. The Alabama Department of Conservation, under the leadership of Commissioner Claude Kelly and Chief of Game and Fish, Charles Kelley, deserve a lot of credit for their vision in promulgating deer and turkey to every county in Alabama.   Other states also live-trapped species of wildlife native to their areas and the result has been outstanding from a wildlife standpoint. Biologists estimate that there are more deer and turkey in America than were present in 1492 when Columbus discovered this new land. This is a real tribute to the American sportsman and the supporting governmental agencies that have labored in the interest of wildlife.

Additionally in 1964, the Alabama Department of Conservation live-trapped and released about 25 turkey on both the Jackson and Marshall County farms. These turkey were baited and trapped with a cannon net on the Fred T. Stimpson wildlife sanctuary. This tract of land is in Clarke County and was donated to the State of Alabama by the Stimpson family from Mobile, Alabama. The cannon net which fires a projectile carrying a net over the flock of turkey is fired from a blind overlooking the baited areas. The flopping turkeys are then removed from the net, put in pasteboard boxes and transported immediately to the release site. There are other methods to live-trap turkey such as individual traps and drug laced feed, but the cannon net proved to be the most efficient method and has the best survival rate. We were primarily still in the cattle business at the time of the turkey release and these turkey transplants did not succeed very well. The Jackson County turkey walked north to Lincoln County, Tennessee and proceeded to reproduce in abundance. The Marshall County turkey walked west to the conflux of the Paint Rock and Tennessee Rivers and set up housekeeping. In the intervening 40 years since the release, the turkey population has grown enough that offspring from the original released turkey have returned to the farms. Our agricultural practices have also changed with less cattle and more row crops and timber management. This provides better nesting habitat which is very popular with the turkey hen. We have learned that where the hen is so will be the gobbler, which is also true for most species of life.

I suppose our family has always had a strong desire to hunt. Most likely this stems from our Indian ancestry. We know of at least three members on the Jones side of our ancestry to be

American Indian, one of which was the famous Powatan Indian princess named Pocahontas. In any event, the love and desire of hunting, the out-of-doors and wildlife have always existed in our family. In chapter ten (Pudge), I described a very meaningful duck hunt for me as a young 13 year old boy to the Mobile Delta country that established a love for duck hunting that still exists in me today. Also in chapter twelve (Other Tracts of Land), I described some of the hunting areas we developed. Mostly our hunting was centered around duck and quail hunting until the establishment of the deer and turkey in North Alabama. Interestingly some of our best duck hunting was on the farm in Jones Valley. The 1950's and 1960's were golden years in North Alabama as far as the duck and goose populations were concerned. This is no longer the case. The waterfowl population on the nearby Wheeler Wildlife Refuge has dramatically declined in recent years. Most likely the decrease in duck and quail numbers contributed to our family becoming avid turkey hunters and moving away from quail and duck hunting.

I served 15 years as the Ducks Unlimited Chairman for the Huntsville area. Ducks Unlimited is an organization that raises money to flood nesting areas for the "mama" duck. The areas are located primarily in Canada. The Ducks Unlimited experience provided me exposure to conservation officials which eventually led me to being asked to serve on the Alabama Conservation Advisory Board, a position I held for 12 years, serving 4 years as chairman. Through friends and acquaintances on this board, I was further introduced to the south Alabama world of turkey hunting. Turkey hunting in south Alabama is not just a sport; it's a way of life. When the avid turkey hunter is not hunting the great bird he's thinking about him. The Jackson County unit of the farm has been the recipient of many hours of effort to establish the wild turkey. Several wildlife corridors and plantings have been established on this farm for the turkey. Timber cutting and thinning has been carried out as an additional attempt to improve the turkey habitat. Warm season grass fields are being grown which provides nesting habitat for the turkey hen. Currently the turkey population is increasing much to the delight of our family. The turkey and deer management programs that have evolved over the years have resulted in numerous wildlife tours coming to the farm. Raymond has been at the forefront of many of these wildlife improvement programs and conducts most of the timber and wildlife tours. Raymond also currently serves on the Alabama Conservation and Fresh Water Fisheries Board. He follows his great uncle, Dr. Walter B. Jones, who was Alabama's first conservation commissioner, as well as my service on this same board. Dr. Walter, as we affectionately called my uncle, was also very interested in establishing the wild turkey on the farm. He volunteered to "babysit" a load of boxed up turkey that Libby and I brought in from South Alabama late one night. Not wanting to release them in the dark, Dr. Walter slept in an old farmhouse with them. The only thing was that when he woke up about daylight the room was full of turkeys that had escaped. We spent most of the morning trying to rebox these uncooperative wild turkey who were reluctant to be released in this "Yankee land" of North Alabama.

Over the years, G. W. Jones & Sons Farms have hosted literally hundreds of farm tours primarily involving cattle, seed and wildlife. Annually we have local grade school classes visit the farm as well as college age and interested farm groups studying cattle, timber and wildlife. We feel this effort, even though time consuming, is a way that we can give back to these interested visitors a little of what we have learned. One thing that remains consistent among farm people all over the nation is that they readily share their ideas with others. This is particularly valuable in establishing wildlife on a farm. Many hours and much failure can be avoided by the sharing of ideas and procedures from one farm owner to another. Even though different, establishing wildlife on the farms has many similarities to the days of Carl and Ed Jones when they started the original G. W. Jones & Sons Farm in the 1940's. I think they would both be pleased by the promulgation of wildlife on their farms. Farmers who have an abundance of wildlife on their land have another real level of value in that hunting rights have become very lucrative and there are many ways to market these rights. G. W. Jones & Sons Farm has chosen not to lease, sell, or market hunting rights as of this date. Our "in house" hunting family remains large and we have chosen instead to involve the next generation with the farm's wildlife, hunting and its expansion as much as possible.

Our love for the out-of-doors and hunting has been exponentially enhanced by our labor to increase wildlife on these farms. Even though we duck and deer hunt regularly, turkey hunting has moved to the front as the preferred species to pursue. We have become avid turkey hunters and go turkey hunting at every opportunity. Our family usually hunts four states annually each spring. As a table fare, the wild turkey is much superior to the meat of its tame cousin. All of our family, young and old, prefer wild turkey meat to any other. My wife, Libby, has become excellent at cooking them and often she gets requests for her recipe. I have killed five of the six sub-species of wild turkey; the Eastern, the Rio Grande, the Merriam, the Osceola and the Goulds (in Mexico). Several of our family members also have similar harvest records of this grand bird called the wild turkey. Ben Franklin

Family turkey hunting trip to Texas

once promoted the wild turkey, via a motion to Congress, as the avian symbol for our nation. His argument was that the wild turkey was the only candidate that was truly American. The Bald Eagle finally won by two votes and became our nation's symbol. Even though I love and respect the gallant wild turkey, I believe the bald eagle is truly regal as it represents our wonderful country.

Time and space does not adequately permit me to give you, the reader, an in-depth description of the world of turkey hunting. The level of difficulty, the physical requirements, the knowledge of the woods and the calls and equipment required would occupy several books. It is, without a doubt, the most intriguing, frustrating and difficult of all the gun hunting sport endeavors. It's really not just a "sport" but "a way of life". With this in mind, I want to close this chapter of "The Farm" with one of my turkey hunting stories from my book "Southern Turkey Hunting" to give you a little flavor of the turkey hunting experience.

ANGEL IN THE MOONLIGHT

The wild turkey has not been very prevalent in my home area of North Alabama until about the mid 1960's. Almost wiped out during the depression, it was not until the restoration days of the 50's and 60's that the turkey in North Alabama began its comeback. Live trapping by our State Department of Conservation and encouragement from wildlife organizations such as the National Wild Turkey Federation (NWTF) commenced a dramatic comeback of the Eastern Wild Turkey in our state. Many areas all over the United States can claim similar successes, all of which culminate in a turkey population nationwide that is more numerous in total numbers than it was when Columbus discovered America. Most of that increase in population has occurred in the last 50^+ years. During this time it would be hard for another species of wildlife in America to stake claim to such a dramatic increase in population as the wild turkey.

Most of my hunting as a boy occurred in the 50's and 60's without ever having seen a deer or turkey in North Alabama. Ducks, squirrel, dove, rabbit and quail were the total fare, with no thought of even seeing a big game animal. Our family farm in Jackson County, Alabama was one of the areas chosen in Alabama for a deer and turkey release in the 1960's. Releases were made over several years of both deer and turkey. From time to time we heard and read stories of turkey and deer hunts and we never realized that these releases would be so successful. At the time of the writing of this story (2006), deer are so prevalent that they are becoming a nuisance. The State of Alabama allows each hunter, if possible, to legally kill 216 deer annually in some areas of the state in an effort to try and control the population. The turkey population also is expanding rapidly and the State of Alabama has a bag limit of five gobblers per season, one of the nation's most liberal limits.

By the late 70's and early 80's, the turkey release, as well as the deer release, was beginning to bear fruit in North Alabama. Sightings were becoming more numerous each year and the population was progressing toward reaching "huntable numbers". I had hunted the Jackson County farm for turkey and deer several times by the late 70's but with very limited success. The farm supported a large acreage of mountain hardwoods and pastureland. Very little nesting habitat was available for the turkey hen so we had very few numbers until 25 years later. During those 25 years we planted numerous wildlife corridors, did some timber cutting, went into the row crop business and established numerous "patch" plantings of autumn olives, chickasaw plums, sawtooth oaks, and pine plantation areas. Today we have a very prolific population of wild turkey, but at the time of this story, 1979, it was sparse at best.

My wife Libby and I were blessed in 1962 with a baby girl with blonde hair and bright eyes. We named her Lisa and she came here wanting to please people and could do just about anything. By the time of the occurrence of this story, she was 16 years old and was very adept with a shotgun. She was the best young "duck shot" of the many young people I have helped train over the years. She could usually make the boy cousins very jealous and upset because she could out shoot them almost every time. Once we went on an ocean cruise and the ship featured trap shooting off the stern. She was only about 11 at the time but I paid for five shots and convinced the man in charge that she could fire a gun. She broke all five clay pigeons, the crowd cheered, so I bought five more shots of which she immediately dispatched. By now the crowd was getting real interested in this little blonde "Annie Oakley" and they started putting up the money to keep her shooting. I don't know how many more clay's she broke, but I only remember her missing one. The man in charge of the trap shooting was still taking in money for her when we were called to dinner.

With this background in mind, the turkey story at hand began in the spring of 1979 on our Jackson County, Alabama farm. Several times I took Lisa that spring trying to find a gobbler in the sparse turkey population of North Alabama. It was late in the spring season and we had had no luck so it looked like our best bet might be to "wait until next year". One afternoon in late April I was checking on our cattle herd when I saw a gobbler and five hens pecking along the south side of a long hollow in the pasture. I was shocked to see them so I cut off the engine to the truck and watched them until dark. I knew this would be the only turkey in the area and I sure didn't want to spook them. I watched them enter the mountain hardwoods on the way to their roost and planned our strategy for the next morning. I eased away in the dark, driving a long way without lights just to be sure not to alert the attention of the turkey. Lighthearted and encouraged I drove home, about 20 miles, and told Lisa to get her stuff ready because we may yet have a chance to at least hear a turkey before the season closed.

Available good turkey hunting equipment was not prevalent 25 years ago. Our guns were our duck guns and our ammunition was 2 ¾ inch rounds and usually #6 shot. Our camouflage was Army issue coveralls and an Army fatigue cap, both of which were olive drab in color. 3-D camouflage clothing, turkey vests or snake boots, etc. were not available. We wore L.L. Bean upland boots and used an Army issue gas mask satchel in which we carried our calls, gloves and shells. The lack of equipment was especially inadequate on hunts with Lisa because of her beautiful blonde hair that was long and flowing and offered a lot for a turkey to see.

Fortunately, our hunt was on a Saturday morning which didn't conflict with school. We left home around 4 a.m. after eating a bowl of cereal and having some fruit juice. We were both excited because we knew approximately where the turkey was roosted and on this morning the weather was classic. The temperature was in the low 50's, no wind and a brilliant full moon lighting the night. The 20 mile trip just sped by as we excitedly talked about and anticipated our prospective turkey. We did several mental checks on what equipment we had and arrived at the farm earlier than I had planned. I drove in the moonlight for the last ½ mile and let the truck roll as silently as possible to a stop near a predetermined fence brace that I had picked out the evening before. Since we were early, I told Lisa that we had about 15 minutes to spare so she could snooze and I would wake her up in plenty of time. She pushed the seat back slightly and went to sleep almost immediately. I didn't close my eyes because they were beholding a beautiful angel sleeping in the moonlight. The Bible says that sometimes we come in contact with angels unaware. Even though its been 30 years since this hunt occurred, I can still see the image of that angel in the moonlight. The 15 minutes sped by and then I awakened my beautiful blonde angel and she bounded up ready for the hunt.

We walked about ¼ mile to our listening spot in the most beautiful setting. The moon was bright, the dew glistened in the pasture at our feet and the sun was just brightening the eastern sky. When David wrote Psalms 19 he must have had a beautiful setting like this in mind. Verse 1 says "The heavens declare the glory of God; and the firmament showeth His handiwork". This was our situation exactly and I had an angel walking beside me in the moonlight. Several owls were busy when we stopped at our listening spot but I went ahead and did my best to mimic them hoping the gobbler would answer something. I also crow called, did my best owling, but we heard nothing until it was getting near daylight. A gobbler sounded off real high in the mountain, too far I felt for us to climb before he left the roost. I told Lisa we had better find another gobbler if we could, so we walked on toward the back of the long hollow but nothing would answer except this same turkey. I realized that he was the only "game in town" so we returned to our original spot, got a good bearing on his position and started climbing.

Lisa was a good athlete, a 3rd base softball whiz at the time, and she kept up on the climb real well. The mountains of Northeast Alabama are part of the Appalachian chain and are steep with limestone rock outcropping and covered in hardwood timber. The turkey heard us walking in the dry leaves and got real excited thinking we were hens responding to his gobble. When hunting these mountain turkey, the hunter should always go to one side or the other of the gobbling turkey and set up on his level or just above him. We did just that and I set Lisa in front of a big oak tree on the lip of a pretty bench just above the roosting turkey. The turkey gobbled incessantly but stayed in the tree. I finally tried to softly coax him down with some clucks and soft tree yelps, but to no avail. He just stayed in the tree and gobbled and gobbled. I whispered to Lisa that we might not kill him but we were sure getting to hear some good gobbling.

The turkey didn't move for about 40 minutes, but kept up constant gobbling and then suddenly he hushed. I thought now maybe he would fly down and come. Sure enough I heard him flap his wings and fly. A long silence went by and I told Lisa to watch closely because he could be coming. I soft yelped, nothing. I yelped again, nothing. I asked Lisa, since she had a better view of the situation, if she had seen anything. She said she glimpsed a turkey fly down the mountain. "Fly down the mountain?" I asked her if she was sure and she said yes. At this point I stood up and yelped with my mouth caller as loud as I could. The turkey answered me from the pasture, almost in the same area as our first listening spot. I told Lisa to pick up her stuff because we were through with this location. The turkey was gobbling profusely now from his established spot in the pasture. I figured our only chance was to walk away from him toward the back of the hollow. We made as much noise as we could walking in the dry leaves hoping he would hear us. Hear us he did, and it seemed to insult him that any "hen" would walk away from his beautiful gobbling. He was gobbling almost non stop by the time we reached the fence at the pasture edge. I sat Lisa up just behind the fence hoping that she would have a good shot in the pasture. We didn't have to wonder where he was because he was loudly making his presence known.

I yelped with the only callers I had at the time, my mouth caller and my cedar box, and did all the tricks I knew. I didn't have to wait for an answer and sometimes I got several answers. This went on for at least an hour, but the gobbler never moved. I told Lisa that we were going to have to change our tactics if we were ever going to fool this crazy turkey. I told her to stay put where she was and I was going to retreat deeper into the cove and maybe I could draw the turkey by her. I left via a nearby dry creek and quickly moved to a spot about 125 yards from Lisa. I eased out to the edge of the woods and saw our gobbler still strutting and gobbling in his same tracks. When I yelped he came unglued and put out some fierce gobbles and started towards Lisa. Now Lisa was between me and the gobbler and he was coming on a straight line for her. I relaxed and thought what a vantage point I had, all I've got to do is watch the show.

To my astonishment though, when the turkey was within 75-100 yards of Lisa he did a circle movement to her right, crossed a ditch just out of her range and made a bee line for me. "Oh no" I thought when I realized that if he got any closer he would see me. The place where I was sitting was a full stand of ankle high poison ivy. I'm very allergic to poison ivy and only walk carefully through it when I have no other choice. It's one thing to walk through it but still another to roll in it on the way back to the ditch so as not to be seen by the turkey. When I was about half way through the roll of the 20 or so feet of poison ivy, I said to myself "Why does it have to be poison ivy?" My situation was similar to the movie scene in "Raiders of the Lost Ark" when Indiana Jones fell into an ancient tomb that was full of snakes and his only comment was "why does it have to be snakes?"

When I reached the dry creek I had cover and I could stand up without being seen by the turkey. Since the late spring vegetation was heavy, I quickly moved back toward Lisa and all the way around her to a point where she was once again between me and the turkey. The turkey was not gobbling as much now and I hoped he had not seen me move. I immediately started yelping on my box call and he would answer, but not as readily as before. I was afraid we were losing him so I tried one last desperate tactic. I cackled with my mouth call and gobbled with my Lynch box. This infuriated the turkey so I gobbled again and he went into a fever pitch. Another gobble and I saw him coming. His demeanor had changed from seeking companionship to fighting this intruder of his territory. Lisa also saw him coming but she and her gun were facing the wrong direction. I could see her slowly turning her gun around in order to get a shot but the turkey was right on her. I gobbled again on my box to distract his attention from her and this made him even madder. Through the trees he looked like a thick black stump entwined with a red Bougainvillea on top and itching for a fight. When he was about 18 steps from Lisa I called for her to "shoot" at which time she made a perfect swing to her right and dropped the turkey cleanly in the field without the turkey ever even seeing her.

Now Superman is supposed to leap tall buildings in a single bound and that is what I did to the fence and I was standing over the turkey in seconds. I pinned him to the ground with my foot, but there was really no need because Lisa had made a perfect shot. Quickly Lisa crossed the fence and reached the scene of the fallen turkey. The cove was quiet except for the whooping and dancing around the turkey by two tired turkey hunters one of which was a very happy blonde haired angel. The hunt had lasted about 4 ½ hours, involved a lot of climbing and walking and there were thousands of gobbles involved. Even that late in the day the wind was not blowing and it couldn't have been a more perfect spring morning.

When the jubilation subsided we relived the hunt and we had a prayer together thanking God for the beautiful morning, the spring foliage, the turkey and this great nation in which we live. The walk back to the truck which was about ¾ of a mile but it seemed like only a few yards. Lisa carried the turkey and I carried the guns and other equipment and then we showed the turkey all over town. The turkey weighed 18 lbs. and had two beards, one nine inches long and the other six inches long and he had 1 1/8" spurs. The story was told to all that would listen then and to all that would listen in years to come. The story with a picture of Lisa and the turkey was printed in our local newspaper and in a statewide magazine publication. I've had many turkey hunts and duels since this story, of which I've won some and lost many, but I shall never forget this particular hunt or the privilege of getting to see an angel sleeping in the moonlight.

Epilogue

Upon completion of High School, Lisa attended Lipscomb and Auburn Universities, receiving a degree in civil engineering. She met and married Mark Yokley, also a civil engineer, and today (2006) they are both registered engineers in the State of Alabama.   Mark and Lisa have two teenage daughters, Elizabeth and Allison, who are also blonde angels and can shoot almost as good as their mother could at that age. Lisa, whom we sometimes call "bright eyes", continues to make life pleasant for all that know her and will always be my "angel in the moonlight".

Ray B. Jones

Huntsville, Alabama

September1998

Chapter Twenty Six – Recognitions

The farm business that Carl and Ed Jones started in 1939 was reaching a stage of maturity by the close of the 20th century. This stage in our history was being recognized locally as well as nationally. We were host to numerous agricultural tours from all over the nation. We hosted college tours, county agricultural groups from Alabama and other states as well as many local kindergarten and grade school students. One of the most meaningful tours we had during this time was sponsored by the Western Livestock Journal. Our farm was chosen by the W.L.J. to host a noon meal and farm tour for selected agricultural leaders, primarily from the western part of the nation. We had a band play for them during lunch, gave them the grand tour of our cattle and facilities and they seemed to have enjoyed the occasion very much.

No small amount of recognition came when G. W. Jones & Sons Farm was selected for the Environmental Stewardship Award in 1995. The Alabama Cattleman's Association submitted the farm to the National Cattleman's Association for this award. Subsequently, we cleared several levels of competition and the farm was selected to represent District II (Southeastern United States) at the national competition in San Antonio, Texas. The NCA sent photographers to the farm for a media presentation that was eventually shown all over the country. We were in most of the nation's agricultural publications and Raymond and I were flown to Denver for a media seminar. We didn't win the national award but we met a lot of really fine people from all over the country. I have found that cattle producers and ranchers are mostly the same from all over the nation. Cattlemen are usually hard working, conservative, church going family people who want only to produce beef. They could care less about government subsidies, which the cattle industry has repeatedly rejected, because their main desire is to be left alone as they manage their land. The Environmental Stewardship Awards program is still active today and fosters a lot of good will for the cattle industry. Interestingly, our farm was the first Alabama farm to win the ESA and represent District II. Fortunately one of the benefits of the program was that Raymond and I completed a course that taught us how to deal with media based interviews. This came in handy in the years ahead when reporters came calling seeking interviews about the farm as well as non-farm issues. On the heels of the ESA, the farm received another significant honor. Our farm was named The Farm of Distinction by the Farm–City Committee of Alabama in the spring of 1996. We were successful over nine other Alabama farms. The Farm–City Committee erected a large sign on the Jones Valley Farm which remained for a year designating it as Alabama's Farm of Distinction for 1996.

We didn't know it at the time but this award automatically entered the farm for consideration for the Lancaster/Sunbelt Expo Southeastern Farmer of the Year Award which is held annually in Moultrie, Georgia. The Ag Expo is the largest agricultural event held in the Southeastern United States. There were eight states represented in the "Farmer of the Year" competition, namely – Alabama, Georgia, Florida, North Carolina, South Carolina, Tennessee, Mississippi and Virginia. The Ag Expo Committee and the Lancaster Company are the primary sponsors of this annual award. They treated all of us representing the selected farms royally. Several dinners were held in our honor and we all enjoyed this special event. Agricultural equipment demonstrations, horse shows, airplanes flying on ethanol, rodeo competition and other livestock events are held annually at the expo. New farm equipment is on display, an antique tractor parade, barbecue cook off contests, horseshoe pitching contests and the list goes on at the expo. It is truly an extravaganza for those interested in agriculture.

The final event for the "Farmer of the Year" award was held in a large open air structure that was once an airplane hangar. The entire expo is held on a large tract of land that was used in WWII times as an airfield for military aircraft. During the events leading up to the final selection, I had become particularly impressed with the farmer from Florida and felt that he would easily win. Each of the eight candidates was visited several weeks prior to the event by three judges flown to their respective farms by Lancaster. From these half day visits with the farmer and his family on the farm, the judges made their decision as to who would be the "Farmer of the Year" for 1996. Libby, Raymond, Kristy and I made the trip to Moultrie and we left just being proud to represent Alabama. During the final banquet, the Lancaster master of ceremonies, Bobby Batson, carefully read the names and a short resume of each of the eight farms. I watched the man from Florida to see his reaction when his name was called. I was so sure he was going to win that I almost didn't hear it when Batson announced Alabama and my name as the winner. They gave us a check for $10,000.00, a new Nissan pick-up truck, a year's supply of Dickie work clothes and a bunch of other things. One of the hardest speeches to give in public is a response that is given after one has received a presentation.

Fortunately, I remembered to thank everyone both present and those back home. It was a great moment for all of the G. W. Jones & Sons farm family. We were introduced all over the expo grounds, asked to judge a barbeque contest, given subscriptions to several magazines and it was a memorable and humbling experience that we shall never forget. G. W. Jones and Sons Farm was the first and still the only Alabama farm to be selected for this "Farmer of the Year" award.

The 1996 Lancaster/Sunbelt Expo Southeastern Farmer of the Year Award.

Raymond and Kristy drove the new truck home after the expo was over. One side note to this event was that Raymond received a letter by mistake about a week before the Lancaster/Sunbelt Ag Expo trip that announced to the Progressive Farmer magazine that G. W. Jones & Sons Farm was the winner. He called them and they swore him to secrecy and he and Kristy kept the secret until after the event. Had I known the outcome, it would have made my acceptance speech much easier. Kristy and Raymond felt that keeping the secret was their choice and they laughed about it for days. Subsequently, we appeared in and on the cover of the Progressive Farmer magazine, and several other agricultural publications. The following year we had over 1,600 visitors to visit the farm in Jones Valley which had certainly come a long way from the one Carl and Ed had started in Garth's Hollow in 1939. Numerous other honors and recognitions have continued to come to G. W. Jones & Sons Farm since that time. The Jackson County Farm was named an Alabama Treasured Forest in 2000. In 2009, the Jackson County farm received the prestigious "Helen B. Mosley Memorial Treasure Forest" award. All three farms have been selected as a host location for "farm field days". These field days are held annually by county Cattlemen's Associations and usually include a meal, farm tour, instruction from Auburn University personnel and a lot of fellowship. Our grandchildren, Elizabeth and Allison Yokley, Caroline, Will and Bryant Patterson, Raymond B. Jones, III and Katherine Jones have been active in the 4-H Club. They have shown heifers selected from our herd in these Alabama 4H shows. Fortunately, they showed the State of Alabama Grand Champion in 2004, 2005, 2006 and 2009 as well as the Reserve Grand Champion in 2005, 2007 and 2008. These 4-H shows are a lot of fun and hard work for the kids and it offers a good way to exhibit some of our cattle herd on a state wide basis.

The farm has been very popular with Huntsville's citizens over the years. People like to see the cattle, especially in the fall when we are calving. Huntsville's growth has had its impact on the farm and in some ways it has been restrictive. In the early days we easily crossed the city streets and roads that dissected the farm. As Huntsville grew, this became more difficult. An underpass under Carl T. Jones Drive served as a way to cross that road from the North to the South and vice versa. Some areas that were once part of the farm were completely cut off by roads with no surface access. Today most of the routes of travel for our cattle crossing the roads have been abandoned. Only one remains today. The crossing of Garth Road occurs several times a year which requires flagmen to stop the traffic, a feed towing wagon and several pick-up trucks driving the cattle across the road. Most of the traveling public enjoy seeing the cattle drive which can be done in about 5–10 minutes. Several times during the year people call and are very helpful when cattle get out. I remember one rainy December morning about 4:30 a.m. when I was going duck hunting that I found two elderly women on foot trying to contain a calf that was loose in Garth Road. I stopped and opened a gate and blocked one side of the "cattle drive" and the calf went back in the pasture. I thanked the two women and they said they really enjoyed "herding cattle" in the rain. I asked them if they needed a job because we could always use some more "cow punchers". They laughed and said they believed not but they sure would enjoy it. The cattle business has been very compatible with Huntsville's urban sprawl. I often get questions from people engaged in agriculture living in other states as to how we can still farm in the midst of a city. The answer is that it has been very compatible because people like to see the cattle and subsequently they feel a part of the farm. The farm, at the time this book is being written, is still considered to be the largest active urban farm in America. This is a tribute to all those who have worked on and for the farm over the last three quarters of a century. I think brothers Ed and Carl would be very pleased with each accolade and recognition that their fledgling 1939 farm enterprise has received in the subsequent years.

Several other recognitions came to our family that were personal and business related during this time frame. Mark Yokley entered the 21st century as a wonderful president of our G. W. Jones & Sons Civil Engineering business. I had run the engineering business since my father's death in 1967. I held it together and over some rough spots but Mark has added a level of quiet, professional leadership that is superior to mine. The firm was inducted into the "Alabama Engineering Hall of Fame" in 2007. Most engineering employees and family journeyed to Birmingham via a bus to receive this prestigious award which was a significant milestone in our business. Mark and Lisa were also involved along with their girls, Elizabeth and Allison, in the farm's 4-H club activities as was mentioned earlier. Mark and Lisa and family were selected as the "Farm Family of the Year in 2007" and Allison was the "2007 Junior Cattleperson of the Year". Currently, Mark serves on the Lipscomb University Board of Trustees, the Huntsville Hospital Foundation and also as an Elder at the Mayfair Church of Christ.

Mike Patterson is serving G. W. Jones & Sons Consulting Engineers as its Chief Financial Officer and has put the flow of the firm's financial information on a very comfortable level. Mike is serving on the Huntsville Hospital Board as well as the Board of Compass Bank. Mike also handles the farm financials and the timely reports of accurate debit and credit information that has added to the stability of the farm.

Our son Raymond, in addition to his farm recognitions, currently serves on the State of Alabama Conservation Advisory Board. He and Kristy helped start the Huntsville chapter of the Juvenile Diabetes Research Foundation. Raymond also serves as the 4-H Club District Show Chairman and serves as a deacon at the Mayfair Church of Christ. All of our immediate family, fifteen total, attends the Church of Christ which Libby and I consider to be a real blessing.

Recognition for Libby and me, aside from the farm recognitions mentioned earlier, also humbled us during this period of time. In 1996, I was elected Chairman of the Board of the University of Alabama Huntsville Foundation, a position I still hold today. In 1999, UAH presented me with an Honorary Doctorate Degree. Libby, the school teacher, had a scholarship named the "Elizabeth M. Jones English Scholarship" that UAH established in her name in 2005. One interesting story happened as I was awarded the doctorate degree from UAH. There were two honorary doctorate degrees awarded at that particular graduation. Pianist Van Cliborne was the other recipient so we sat together on the front row of the stage. Van Cliborne is a world renowned pianist having won piano competitions all over the world. He was raised in Texas on a farm so we had a lot to talk about during the ceremonies. Afterward, my sister Betsy was startled and very surprised that I could even converse with such an accomplished man as this classical pianist. "Ray, what in the world could you possibly talk to Van Cliborne about?" she asked with excitement. I said, "Betsy we talked about music." She put her hands on her hips, much like she did when she was a little girl, and said, "Ray, you don't know anything about music". I said, "Betsy all I had to do was ask him if he knew the song "Sleeping at the Foot of the Bed" and the conversation went downhill from that point." With this she flounced off and I didn't tell her any different for quite a while.

In 2002, I received the Huntsville Distinguished Service Award from the Huntsville Madison County Chamber of Commerce, an award my father had received in 1965. It was the first father / son award in its history so the award was very special to me as well as the whole family. Also in 2002, Lipscomb University named its School of Engineering the "Raymond B. Jones School of Engineering." This was very humbling to me also since I was trained as a farmer and not as an engineer.

2007 Alabama Business Hall of Fame

In 2007, I was inducted into the Alabama Business Hall of Fame by the University of Alabama in Tuscaloosa. Family and several friends made the trip to Tuscaloosa via a private bus for us to receive the award. Interestingly, my father was inducted posthumously into the Alabama Business Hall of Fame in 1983 which also made this one very special. In 2005, I retired from the Regions Bank Board after 38 years service as the longest tenured director in their system which is still true today. In 2006, I retired from Lipscomb University's board after 23 years service. The most recent recognition I received was being inducted in the Junior Achievement of North Alabama Business Hall of Fame in 2009. I was honored to be recognized in the company of several other great Huntsvillians.

Libby and I are most grateful for all these blessings and others that have come to our family. We give God the credit and count it a special privilege to be recognized also as part of His family.

Being born an American, raised on a farm in the South, being blessed with a wonderful family, marrying an absolute angel and working with some great people over the years in tilling the soil and seeing God's miracles being performed each day has made me a most blessed and humbled author of this book.

Chapter Twenty Seven – Changing Valley

Over the last 70 years I've seen the farm and Jones Valley change several times. Most changes in life are gradual, particularly when one is living near, involved in or around the changes. I suppose everyone has at least some "built in" resistance to change. We are usually comfortable with our surroundings and way of life. Change causes us to long for the way things were, or at least the way they were perceived to have been. Someone has said that "change is inevitable" which has certainly been true of the farm and Jones Valley as well.

Change was slow for the farm and Jones Valley in the early years. WWII had a dampening affect on the growth and change of the whole county. The ending of the war and the survival of Betty and her family was a significant milestone in the history of the farm. The metamorphosis involving the return of Ed and Carl from the war and replacing a row crop farm with a cattle and seed enterprise was significant. Obviously, the influx of new people to Huntsville, and subsequently many of them moving to the valley, made for a changing valley. The valley was fast changing from cotton and livestock farms into a residential community. The Hill family was one of the first families to subdivide their farm (Bel-Aire Estates) closely followed by the Acuff/Wilmer development (Greenwyche). The Hill land had been row cropped and was south of the farm, while our northern neighbors, the Acuffs, had mostly livestock consisting of dairy cattle and sheep. In rapid succession, others followed this subdivision trend which brought many new families to Jones Valley. The 1950's and 1960's saw this development which had begun in the north part of the valley move steadily south towards the Tennessee River, forever changing the valley from agriculture and timber to a city accompanying a growing population.

The changing valley imposed a demand on the farm even though the Jones family, unlike our neighbors, chose at that time not to develop or subdivide its land. Rights-of-way for Garth and Four Mile Post Roads, a sewer line along Aldridge Creek and several electrical and water easements were requested by the City to satisfy the growing population. The farm responded to the demands of this growth from the surrounding development by donating to the City, free of charge, the land for these rights-of-way and utility easements. During these years, the City Board of Education was adding a school room a week and had a critical need for an elementary school site. They approached the Jones family about purchasing a site on the farm. Instead of selling a tract of land for a school, the family donated the present Jones Valley School site to the Huntsville City School System.

In the 1960's, residents began pouring into the City at an alarming rate and this continued through the 1970's. Huntsville was listed as the fastest growing city in the nation for many of those years. The town was known as a family town, had more churches per capita than any city in the nation and was known as a conservative, wholesome place in which to live and raise a family. Memorial Parkway was finally finished and commercial establishments moved to this new by-pass. Parkway Shopping Center became a favorite, as did other places of commerce along the Parkway. Conversely, with all this phenomenal growth in the City, the farm remained essentially the same, continuing to raise cattle and seed and carry on the agricultural practices started by Carl and

Ed Jones in 1939. Still, pressing needs of the City continued to be made on the farm to accommodate the demands of an ever growing and expanding city. A new road connecting Bailey Cove Road and Whitesburg Drive was envisioned. This new road, which became known as Carl T. Jones Drive, was to dissect the farm and take 27 acres of land. Again, the family donated the land to the City for this new road.

Some development of the farm was precipitated by the building of Carl T. Jones Drive, such as Jones Valley Gardens Phases I and II, Somerby at Jones Farm, Southwood Presbyterian Church, Mayfair Church of Christ and the Valley Bend Shopping Center. One would not have to look very closely at the architecture and landscaping of these developments to realize that it is the desire and intent of the Jones family to see that any changes to this valley will be of quality and beauty. This has been our home since 1939 and hopefully our family will be here for many decades to come in this beautiful place; a farm that our family has nurtured and referred to over the years as "God's Country" and a place we have chosen to preserve and protect. Even though changes will inevitably occur in the valley and on the farm, it is the Jones family's present plan to continue to pasture and raise cattle.

The "Home Place" in Jones Valley is part of us, where sweat, blood and tears have literally been involved in its making. The intent of the family to pursue agriculture in the valley is both obvious and important. The past three or four decades have seen our neighbors develop their property and mostly move away from the valley while our family has stayed and remained in the agricultural business that was started 70 years ago. Each time a development on the farm does occur, it is filled with emotion for me. On a still summer day, even after all these years, I sometimes can imagine that I hear the click of a mule's shoe against the trace chain or the dinner bell ringing at the big house signaling to those in the fields that it was dinner time. Sometimes on a summer night when I take a walk, I can still hear in my mind the windrowers swathing the grass seed with all the tractors running in unison and the men singing at the top of their voices all night long. To this day, the lowing of the cattle reminds me of yesteryear when there were many more cattle and also of the wonderful people who are now gone who tended them. These thoughts and emotions I realize are "gone with the wind" and eventually the farm and the valley will change even more. Our family's desire is that whatever we develop of the farm in the future will be done in good taste and in a way that would make Carl and Ed proud. This piece of land that they purchased and nurtured in the valley has meant so much to thousands of people over the years. It is our family's obligation and legacy to guide this farm into the future in such a way that would be pleasing to all those who have been connected with it. Change, I'm sure, will inevitably come. When it does, it's my prayer that we will pass this blessed land on to the next generation invoking them to nurture it like our family has done for almost three quarters of a century in this wonderful country that we call America.

The Changing Valley

EPILOGUE

Once upon a time, there was a small farming community in a beautiful valley just southeast of Huntsville, Alabama. The citizens of this little 'Camelot' all labored to mold this green farm into the dream of two brothers. With much labor and willing hearts, their efforts were successful and the farm has lasted almost three quarters of a century. People, young and old, black and white, big and small, have poured their lives into this farm and have loved the land while never doubting their purpose.

Each subject of this little 'Kingdom' working on this farming endeavor was somehow connected to the founders of the farm. Seldom in any generation does a farm or a community have such patriarchs as Carl and Betty Jones to set such admirable standards for its direction. Integrity, honesty, hard work and vision seem to still penetrate the thinking of subsequent generations of family as well as laborers who tread the fields of the farm. It is my earnest prayer that this lofty flame of characteristics will never die out of our family. I pray that God will continue to bless this piece of land that we have affectionately called "The Home Place" and all those who have loved it over the years.

Raymond B. Jones